# *HARLAN'S GLOBETROTTERS*

---

## The Story of an Eclipse

by

DAVID S. EVANS AND KAREN I. WINGET

**To order additional copies of this book, contact:**
Xlibris Corporation
1-888-795-4274
www.Xlibris.com
Orders@Xlibris.com
20110

For

John Edwin  Elliott

who gave the team its name

and the late

Harlan James Smith

who coached it.

# The Team Roster

This list duplicates that in the first report
published in the Astronomical Journal
**81**, 452, 1976

R. Allen Brune Jr., Charles L. Cobb, Bryce S. DeWitt, Cécile DeWitt-Morette, David S. Evans, Johnnie E. Floyd, Burton F. Jones, Raymond V. Lazenby, Maurice Marin, Richard A. Matzner, Alfred H. Mikesell, Marjorie R. Mikesell, Richard I. Mitchell, Michael P. Ryan, Harlan J. Smith, Alassane Sy, and Charles D. Thompson.

## Acknowledgements

We also repeat the list of acknowledgements included in that publication, as the original progenitors of the whole enterprise.

The National Science Foundation, and especially Dr. R. La Count. Dr. G.W. Curtis, and Dr. N. Medrud Jr. of NCAR . To the Research Corporation of America, to the University of Texas at Austin, to NATO and the National Geographic Society. To Professors R. Michard and J. Texereau of the Paris Observatory, and to Dr. Margaret Burbidge and Mr. C.A. Murray of the Royal Greenwich Observatory, England. To Professors R. Dicke, D.T. Wilkinson and P. Crane of Princeton University, and to Dr. S. Vasilevskis of Lick, Dr. J.H. Baker of

Harvard, Mr. W.C. Miller of the Hale Observatories, Dr. R. S. Harrington of the U.S. Naval Observatory, Professor J. Ehlers of the Max Planck Institute In Munich, to Mr. E.J. Hahn and Mr. A.G. Millikan of the Special Products and Research Division of Eastman Kodak. The Oxford University press. To all of them we express our warmest thanks.

Especial thanks were also expressed to the Government of the Islamic Republic of Mauritania, and for generous help and hospitality, both official and private, to Dr. and Mrs. Ba Bocar Alpha of Nouakchott, Governor Yarba Ould Ely Bourka of Atar, and Mr. Mahmoud Ould Amar Cheine also of Atar, Préfet Kane Abdoul Mame N'Diack of Chinguetti, also to Mr. A. Akoni of Air Afrique, Mr. P. LeBris of Humbert Travel Service, New York, and Mr.H.H. Benjamin of Braniff International Airways.

It would not have been possible to produce the present text without the generous help of the surviving Globetrotters. Regretfully Richard Mitchell, Charles Thompson, and Alassane Sy and, of course, Harlan Smith, are no longer with us. Professor Bryce De Witt has very generously made available to us the detailed texts of the journals he kept during his West African tour of site selection, and at the actual eclipse expedition. We have made full use of these documents, unchanged except for some editing, with facts always sacrosanct, designed to make them read more easily as a past record of events. They include, of course, much of the special details referring to his wife, Professor Cécile DeWitt-Morette. Charles Cobb and Professor Burton Jones have kindly sent us their complete eclipse files and some additional narrative material. Professor Richard Matzner has responded cheerfully, as have the forementioned, to merciless grilling designed to revive memories and reveal forgotten details. Alfred and Marjorie Mikesell have been assaulted through the ordinary post, and have always responded in friendly and informative manner. The massive collection of documents in Austin attest in no uncertain fashion to the Herculean efforts which he exerted to get the whole enterprise working. To all of these we express our most sincere thanks, as we do also to Johnnie Floyd, chief engineer at the Austin Department at the time, and to Dr. Fritz Benedict for details of the PDS measuring machine. Professor J.P. James of the University of Manchester, England, gave us a good account of the British proposals for an expedition to Niger.

For the account of the follow-up expedition, the reader must rely mostly on the recollections of the senior author, who apologizes for any defects of his senescent memory. illustrative material, other than that cited as not originating from a Texas source, comes from instruction manuals prepared for the use of the expedition members, and from personal photographs taken by the DeWitts and the senior author. Regrettably most of the former material has been lost.

## A Note on Currencies

All finance relating to grants and expenditures in the USA was conducted in US dollars ( $ ). The grant from NATO was paid in Belgian Francs ( FB ). Burton Jones' expenses incurred when he was based in Britain, were rendered in British Pounds ( £ ). The DeWitts, especially Cécile, who was still a French citizen, had dealings in French Francs ( FF ). For a time, the former French-ruled territories of West Africa continued to use local Francs ( CFA = Communité Financier Africain ), but, on June 30 1973, the Government of Mauritania converted to a new local currency, the Ougiya ( UM ).

Approximate conversion rates at the time of the eclipse were:—

$ 1.00 = FB 40 = FF 4 = CFA 200 = UM 40 = £ 0.4

# Acronyms

| | |
|---|---|
| NSF | National Science Foundation |
| NCAR | National Center for Atmospheric Research |
| NGS | National Geographic Society |
| NATO | North Atlantic Treaty Organisation |
| CNRS | Centre National de Recherches Scientifiques |
| ONR | Office of Naval research |
| IAU | International Astronomical Union |
| SOMIMA | Société Minière de Mauritanie |

## Photo Credits

Dr. William Pence, Mr. A.J. Boule, University of Texas,
Royal Astronomical Society, Dr. John James,
Mr. Charles Cobb, Dr G. W. Van Citters.
If not otherwise cited, photographs are by Bryce or Cécile
DeWitt, or by David S. Evans.

## Graphics

Banaby H.W. Evans or reproduced from Evans, 1952

# CONTENTS

# List of Figures

Harlan James Smith, ( 1924-1991)
Chairman, University of Texas Astronomy
Department,
and Director of McDonald Observatory.
(Photo. University of Texas)

# *HARLAN'S GLOBETROTTERS*

The Story of an Eclipse

by

DAVID S. EVANS AND KAREN I. WINGET

# Raison d'Être

Yes indeed, an explanation is called for. The year 1999 in which this writing is being commenced, likewise the successive years, into which it may spill, are drenched in nostalgia. The particular form which it took in our case is that the senior of the two authors, both of whom are associated with the Department of Astronomy and the McDonald Observatory of the University of Texas at Austin, reflected that quite a number of important enterprises had been undertaken over the last half century by this dual organization, of which no proper account had been kept. This prompted the emergence from deepest storage of large boxes of records, which demanded the imposition of some sort of order, the execution of trivial and duplicate documents, and their organization into a form suitable for deposition in the official archives of the State of Texas. The junior author offered to help and was immediately saddled with all kinds of investigatory tasks. The trouble was that we started reading the documents, and found them fascinating, especially those connected with the expedition sent by the University of Texas, in collaboration with others, to observe the great total solar eclipse of 1973 from a site in the oasis of Chinguetti in the Saharan desert republic of Mauritania. As it happens, thirty years later, all the participants save three, are still alive and have been interrogated to record their impressions. The principal exception is Harlan J. Smith, at the time both

Chairman of the Department of Astronomy and Director of the McDonald Observatory—we were much smaller then—who died in 1991. So we have determined to write a book about the whole affair, and it only seems right to give it the title coined by one of the senior author's students, John Elliott, "Harlan's Globetrotters"—in imitation of the celebrated basketball team, with a subtitle to show what we are really talking about.

Well, what's so marvelous about an eclipse expedition? Isn't it all written up in the final published paper? Certainly, if you like bland prose, and the omission of all the things, some ingenious, some ludicrous, which happened to all those concerned in this enterprise. Remember: this was the epoch of jumbo, but not mastodon, jets, when airlines served edible food, when there were still lots of places not overwhelmed by tourists, and when space astronomy had yet to steal ground-based astronomy's thunder. This was a time, when such an expedition, never likely to be repeated, could still be regarded as an adventure.

## Some Eclipse Basics: Shadows

Eclipses occur when the shadow of one celestial object falls on another. In the solar system the light source is the Sun, and for the present purposes the objects of importance are the Earth and the Moon. A lunar eclipse takes place when the shadow of the Earth produced by the light from the Sun, falls on the Moon. This requires the Sun, Earth and Moon to be close to the same straight line, with the Earth in the middle, that is to say, eclipses of the Moon necessarily take place when the Moon is full. Eclipses of the Sun take place when the shadow of the Moon falls on the Earth. Now the three bodies are aligned with the Moon in the middle, so that eclipses of the Sun necessarily take place at new Moon.

The geometry of these situations will be clear from Figure 1, which shows that the shadow of the Earth or Moon cast in the light of the Sun, is a quite complex structure. Because the Sun is so much larger than either of the other two bodies, there is a region of complete darkness extending on the side away from the Sun for only a limited distance. It is in the shape of a circular cone, and comes to an end in a point. This region of total shadow is called the *umbra*. On the figure you will find that you cannot draw an uninterrupted straight line from any point inside the umbra to any point on the Sun's surface. Outside the umbra is a diverging conical volume from any point of which the Sun's disk is partially obscured. The proportion of obscuration ranges from maximum on the axis of this region—the extension of the umbra—to vanishingly small at the surface of this cone. This region of partial obscuration is called the *penumbra*. One rather special case is the appearance of the Sun from a point near the common axis of the umbra and penumbra, lying farther away than the end point of the umbra. Seen from any such point the Sun is at most partially obscured leaving free a ring of light, which is the greater or the less, the farther back the observer goes. Such a ring of illumination of the Sun's disk is given a name of Latin origin, meaning a little ring,—an *annulus*—and an eclipse producing this type of situation at maximum obscuration is called an *annular eclipse*. The senior author spent many years of his young life confusing the similar words, *annulus* and *annular*, under the wrong impression that annular eclipses occur every year—they don't.

## Primitive Reactions to Eclipses

Eclipses are not very rare events when considered on the global scale, but in ancient or isolated times, the occurrence of an eclipse, whether of the Sun or the Moon, by accident seen from a particular place,was often regarded as a semi-miraculous event, reflecting either something phenomenal which had occurred in terrestrial history, or presaging some dire catastrophe.

Accidental observation of lunar eclipses is not particularly rare since the phenomenon is that of the incidence of a shadow on the Moon's surface, and thus is visible throughout a large area of the Earth's surface from which the full Moon is seen above the horizon. When a lunar eclipse is total, that is, when the Moon passes entirely through the umbra of the Earth's shadow, the only illumination falling on the Moon's surface is scattered sunlight which has passed through the Earth's atmosphere for many miles, in paths almost tangential to the surface, and has suffered, as the light of the setting Sun does, the almost complete absorption and/or scattering of the blue component of normal sunlight. The result is that the totally eclipsed Moon is often illuminated by a deep orange-red glow, which was often interpreted by contemporary soothsayers in such phrases as, 'The Moon was turned to blood'.

When eclipses were recognized as natural phenomena, as they seem to have been in ancient China, India, Babylon and certainly ancient Greece, lunar eclipses compelled the inference that the Earth was round and about twice the size of the Moon, readily deducible from the curvature of the umbra and its dimensions.

In solar eclipses the dimensions of the Moon's umbra are such that its cone does not always reach to the Earth's surface, the orbits of the celestial bodies involved not being perfectly circular, resulting in a range of dimensions of the eclipse diagram. When this happens, the Sun's disk is not fully covered for a terrestrial observer even at maximum obscuration and an annular eclipse results (Figure 2b). In slightly different circumstances the cone can extend to the Earth's surface, but only just, producing a shadow patch from

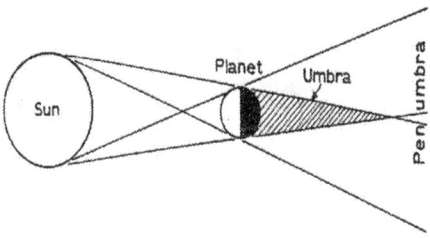

Figure 1: From within the umbra the Sun is totally hidden: from the penumbra, partially so: from the extension of the umbra, an annular eclipse is seen.

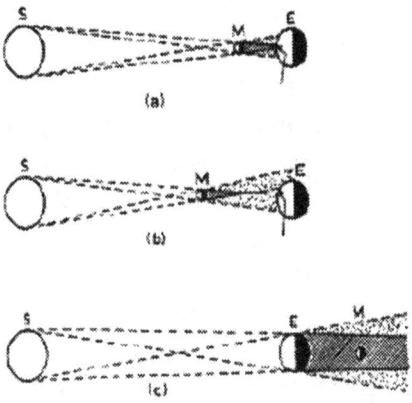

Figure 2: Eclipse circumstances:

(a). The lunar umbra extends to the Earth; the shadow patch moves rapidly over the surface, producing total solar eclipses, each of a few minutes total duration, to be seen only from places over which the umbra passes.
(b). The lunar umbra does not reach the Earth's surface. The solar eclipse is partial or annular.
(c). The Earth's umbra is much larger than the Moon; there is widespread visibility of lunar lasting up to three hours.

within which the eclipse can be seen as total, no more than some 250 miles in diameter, and because the bodies involved are in motion, lasting at any one point, for at most some seven minutes of time (Figure 2a). So, a total solar eclipse can only be seen from within a narrow belt from a position which tracks rapidly across the surface of the Earth. Outside this belt the obscured proportion of the Sun's disk falls off, and partial eclipses will be seen, though fairly close to the totality belt the maximum degree of obscuration may be sufficiently striking to attract the attention of a casual observer.

J.K. Fotheringham, an Oxford don who combined a deep knowledge of ancient manuscripts with a capacity for celestial mechanical computations, sought out instances where events had been linked by ancient chroniclers with contemporary solar eclipses. This was not as simple as it might sound, even with possible sources identified. The author of the chronicle was presumably in most cases, not an eye-witness, might not have been above improving the narrative with a little fiction, and the natural event referred to might not have been an eclipse. But it did raise the possibility that some historical event, whose date and even place might have been uncertain by many years, recorded as it was in a calendric system having little in common with our modern one, might now be identified as to date and place within a matter of minutes and miles simply because the totality zone of a total eclipse is so narrow and any one observer only sees the total phase for so short a time. He was successful in nine cases and so contributed significantly to the chronology of ancient civilizations. But he also found that the computed circumstances of long ago eclipses, found by Victorian era astronomers, with no help in calculation, not even a mere fraction of that available to the modern owner of a ten-dollar pocket calculator, were all off a little systematically, enabling him to determine the long-term rate of slowing down of the Earth's rotation, an effect produced by the friction of the tides in shallow seas.(Fotheringham, 1920)

One feature deducible in ancient times from the fact that the lunar umbra just about reaches to the Earth, would be that the distance to the Moon is about 114 times its diameter, and the distance to the Sun in the same ratio to its diameter.

Lastly we may include in this miscellany a mention of Francis Baily, an early Victorian merchant with a penchant for astronomy, a founder of what became the Royal Astronomical Society, and a considerable contributor to astronomical science, who gave a precise description of the phenomenon now called Baily's Beads (Baily,1838). These are produced at a total solar eclipse when the last rays of the almost hidden crescent shine down rifts and valleys on the lunar limb, producing a brilliant short- lived string of beads of light. (Figure 3). Another striking eclipse phenomenon is produced when the visible Sun is reduced to a mere crescent, at which time all the small gaps in the leaf patterns of trees act as pinholes and produce bright crescent images on the ground and nearby walls. (Figure 4)

# The Expeditionary Era

Even in the nineteenth century, it was tacitly accepted that significant astronomical problems might demand travel to strange places, possibly for relatively lengthy periods. Edmond Halley's expedition to St. Helena in the 17th Century, Nicolas de Lacaille's to South Africa and Mauritius in the 18th, and Sir John Herschel's also to South Africa in the 19th, all for studies of the southern sky, are examples of this. There were also many much shorter ones to all sorts of strange places, as dictated by the astronomical circumstances. These included the enormous international and remarkably coordinated observations of the Transits of Venus in the 18th century, to be repeated for those in the next. In those two centuries the disk of the planet Venus could be seen on two occasions traversing the Sun, a phenomenon which could

Figure 3: A Baily's bead at the 1973 solar eclipse.
(Photo. Dr W. Pence from aboard ship)

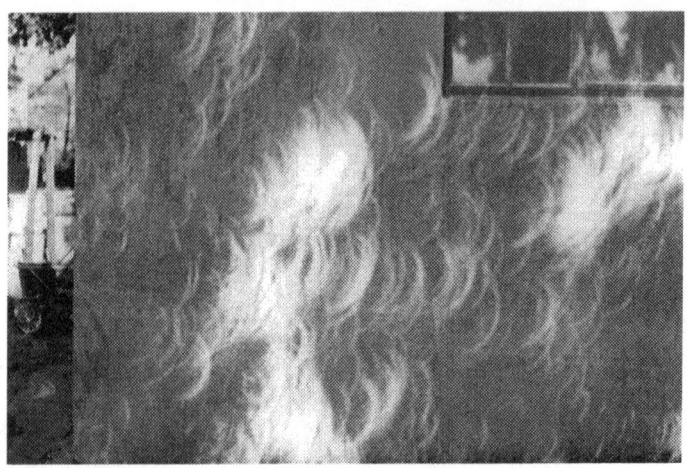

Figure 4: At a near-total annular eclipse,
the crescent Sun was imaged on walls by gaps in
foliage acting as primitive pinhole cameras.
(Photo. A. J. Boulle, Huguenot, South Africa)

be used to determine the distance of the Earth from the Sun. Even so, most of these efforts took a long time and involved extended voyages in sailing ships. Expeditions to observe total solar eclipses became much easier with the development of steam-powered ships and many were undertaken. It would be a mistake to think of them in terms of those which might now be launched by such large official organizations as NASA or the NSF. With the exception of a few relatively modest establishments such as the Royal Observatory at Greenwich, astronomy through most of the nineteenth century, was conducted either by, usually, a two-man teaching faculty from a few universities, or, still more, in Britain by the tradition of the wealthy and benevolent amateur, such as, to name only one, Sir William Huggins, (and his wife).

Most of these expeditions used the burgeoning techniques of spectroscopy aimed at the almost-eclipsed Sun to discover more about its structure. This yielded remarkable discoveries: the extra hot thin layer of gas surrounding the Sun to a height of some thousands of kilometers, the *chromosphere*, which produces not the absorption spectrum of the solar disk, but an emission spectrum. Sometimes rising through the chromosphere are to be seen great arches or plumes of luminous hydrogen, known as *prominences*, (Figure 5), rooted near active sunspot regions. Joseph Norman Lockyer, an amateur pioneer spectroscopist who discovered the chromosphere, found in it an unknown spectrum line which he attributed to a new chemical element, helium, later identified in terrestrial contexts. (Meadows, 1972) It is a measure of the amateur status of contemporary astronomy that Lockyer's day job, as a clerk in the War Office (War Ministry), left him liberty to travel to a number of eclipses, and no less, to found and edit the scientific weekly, 'Nature', still a dominant publication. He was, in the end, rewarded with an institution of his own at the new Imperial College in London.

Figure 5: A solar prominence caught during the photography for the 1919 solar eclipse.
(Photo. Royal Astron, Soc.)

Figure 6: The solar corona at the 1973 eclipse.
(Photo. Dr W. Pence from aboard ship.)

Attention also focussed on a still larger structure surrounding the Sun, the solar *corona*. (Figure 6). Though, of course, always present, it is only readily seen, with a total brightness approximating that of the full Moon, as a white plumed, or arched structure surrounding the eclipsed Sun and extending out to several solar diameters. Special techniques reveal its much greater real extent. The structure of the corona varies frequently, and especially in tune with the general level of solar activity indicated by the sunspot cycle. It may well have long been taken as merely a scattering effect of terrestrial clouds, until comparison of its appearance from two widely separated eclipse observation sites showed that it is indeed a solar appendage. It is a region of enormously high excitation characterized by a temperature parameter in the millions of degrees, and violent ejections of material.

The expeditionary era really ended with the onset of space astronomy offering the possibility of continuous observation instead of a few snatched minutes. There remained one eclipse problem, the subject of this account, to which we turn after a slight detour.

# The Recurrence of Eclipses

Though eclipses only occur close to the new and full phases of the Moon, the reason that they do not occur at every such phase is that the lunar orbit is in a plane tilted with respect to the solar orbit at an angle of some 5 1/2 degrees. At most critical phases the bodies pass each other with the Moon well north or south of the Sun, or its opposite point on the sky.

The Sun moves along a great circle path on the sky which is tilted with respect to the equator, so that the Sun is north of this in northern summer and south in winter. This path is called the *ecliptic* and marks the fundamental plane of the solar system. The Sun covers an annual circuit measured from its crossing point from south to north (the vernal equinox) in a year of 365.2422 days, the so-called *tropical year*, which tracks the phenomena of the terrestrial seasonal cycle.

For an eclipse to occur, say a solar one, the Sun must be near an intersection of the Moon's orbit on the ecliptic, called a *node*, at the same moment as the Moon is also at this point. This situation, the coincidence of Sun and Moon, ensures that the Moon is new. There is a complication, namely, that although the inclination of the Moon's orbit changes little on the average over very long periods of time, the orbit is not fixed but rotates in a period of 18.6 years. The consequence is that the Sun meets our selected node, not in a year, but every 346.62 days. The Moon, of course, gets round its orbit in a month, at an average interval of 29.5306 days. If we started with the Sun and Moon at opposite nodes the Moon would be full, and we should be dealing with a lunar eclipse, for which the arguments are similar. We ask ourselves. If we start from such a coincidence, when can we expect another? This requires a little number juggling, and the simplest answer seems to be the following : nineteen nodal passages occupy, 346.62 x 19 = 6585.78 days and 223 recurrences of full moons require 223 x 29. 5306 days = 6585.32 days. Almost exactly the same number. There is nothing particularly magical about this. As a matter of simple arithmetic one could start with any two numbers whatever and find multiples of each so that the products nearly match. In this case it is remarkable that with relatively small multipliers, the match is so close. So, after the lapse of 6585 days the critical situation will be recreated and there will be another eclipse. This period, amounting to 18 years and 10 or 11 days, (depending how the leap years fall), has been given the inappropriate name, *Saros*, derived from a Babylonian name for a cycle of 3600 years. It is by far the most certain recurrence period of eclipses, whether solar or lunar: but is it the only one?

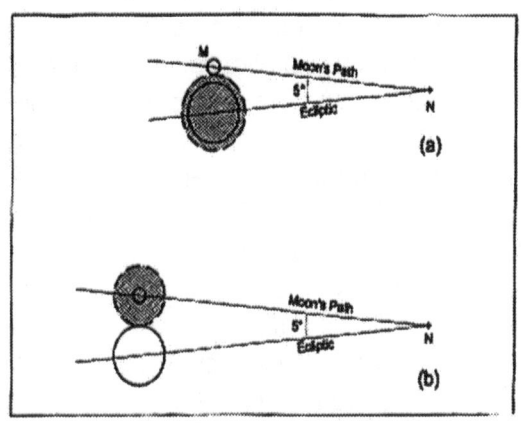

Figure 7: Ecliptic limits:

(a): The Earth's penumbra can just reach the full Moon if this occurs within about 10° of a node, or ten days of the Earth's motion.

(b): The Moon's penumbra can just reach the Earth if new Moon occurs within about 18° of a node, i.e. 18 days of the Eath's motion.

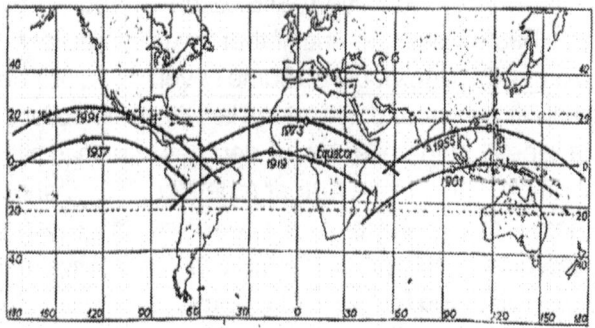

Figure 8: Tracks of the six great solar eclipses of the twentieth century. (Reproduced by permision of the Oxford University Press, from Dyson and Woolley, 1937).

Eclipses can, however, occur with full or new Moon phases some way from lunar nodes. The penumbra of the Earth's shadow, centered on the ecliptic, can reach across, and just touch the lunar disk in its orbit, to produce a minimal partial eclipse, if full Moon occurs no more than 9 ° from a node, as much as 12° in favorable cases—the range is due to the facts that the orbits of Earth and Moon are ellipses, not perfect circles. The limiting condition for a solar eclipse, certainly no more than partial, somewhere on the Earth, requires the lunar penumbra, centered on its orbit, to reach the terrestrial disk on the ecliptic. The limit is that new Moon must take place less than 18° from a node. Since the Sun advances round the ecliptic at just short of 1° per day, the degree numbers may be interpreted as days. (Figure 7). This brings us to the Metonic cycle, discovered by a fifth century BC Athenian astronomer, which is another case of near match of multiples.

Nineteen tropical years come to 6939.6 days, and 235 lunar months come to 6939.7 days, giving the result that if there is, for example, a new Moon on a particular date, there will be another new Moon on the same date 19 years later. If there were a solar eclipse on the first date, there might be another one, because, at least the lunar phase is correct. But the Sun passes through a node every 346.62 days, and 20 of these amount to 6932.4 days, so that if there were an eclipse on the first occasion, the Sun would be some seven days, or 7 degrees off position at the second. This does lie within the ecliptic limits, and at a repetition the offset is 14 degrees, which is still within the limit range, but not for a further repetition. The upshot of this is that there is also a nineteen year periodicity in the occurrence of eclipses, but it cannot persist for more than about five members in the solar case and four in the lunar case.

Eclipse recurrence effects include:

If there is a solar eclipse on a certain date:
There may be a lunar one two weeks later

There may be a solar one four weeks later
There will be a solar one near an opposite node 173 days later
There may be a solar one on the same date 19 years later
but this cannot lead to a series of eclipses of more than five.
There will be a solar one 6585 days later

Because the original eclipse in the above scheme can be replaced by any of the later ones, eclipse recurrence is a complicated overlay of at least two sets of periodicities, and a variety of certainties and uncertainties. The upshot is that the minimum number of eclipses in a year is two, and both are solar. The maximum number possible is seven, with four or five being solar.

At what point was any of this scheme recognized? The fact that the word *Saros* is Babylonian, does not necessarily mean that it was used in its present sense to predict eclipses in ancient times. The fact that the Metonic cycle was known in ancient Greece, does not imply that it was used for eclipse predictions. Its long-time application is in the regulation of the Hebrew calendar and in the computation of the date of Easter.

Evidently there was some early recognition of periodicity, implying that it must have taken place in civilizations having long-term record-keeping, well beyond the presumably very short expectation of life in early times, with a good calendric system, and a number system permitting operations a little beyond the four simplest. One would guess that the key would lie in the nineteen-year cycle, even though in any one instance, this is relatively short-lived, but as against that, it could apply to lunar eclipses observed from a single station, or in certain cases, even solar eclipses for which this may be true. The nineteen year cycle, not always properly understood, is the one favored by interpreters of ancient structures believed to have astronomical significance.

But the recognition of periodicity and even a consequent capacity for prediction, does not imply a capacity for prediction at a particular place, as we shall see. However, Thales of Miletus was credited in Victorian times with the prediction of the eclipse of 585 BC, and the unfortunate ancient Chinese, Hsi and Ho, were executed in the 7th century BC, whether for failure to predict a solar eclipse, or for ignoring one that turned up unexpectedly—the first of a nineteen years series, perhaps—is not clear. (Needham, J. *et al.*, 1960)

Two final remarks: the utility of eclipses for study of the Sun is the result of a remarkable series of coincidences. If the Moon were ten per cent smaller or ten per cent farther away, there would never have been any total solar eclipses, and in the absence of the near perfect fit of the lunar and solar disks, the discovery of the chromosphere would have been much more difficult. If either parameter had been ten per cent larger it would have been even more difficult to detect these features, and even perhaps the corona as well. This fortunate situation is, indeed only temporary, because the dimensions of the lunar orbit are steadily changing, though increasing by only 1.5 inches a year, so that the present situation will last a very long time.

## Six Great Eclipses of the 20th Century

The Saros interval is not a precise whole number of days, which has the consequence that the track of an eclipse predicted by the application of this interval is displaced round the Earth to the east by about 120 degrees of longitude, and the subsequent members of the series by equal further amounts. After three such steps we arrive at a track in a longitude range fairly similar to that of the first eclipse, but displaced north by a few hundred miles. The series of eclipses which we are now considering took place on May 11 1901, May 29 1919, June 8 1937, June 20 1955,

June 30 1973 and July 10 1991. That which took place on June 30 1973 was the twenty-fourth in a series which began with an annular eclipse in the Antarctic on October 11 1558, (Old Style), and will end with the seventieth member in the Arctic. The ones in the twentieth century were all total with immensely long totality tracks, and at maximum, giving totality periods close to seven minutes of time. (Dyson and Woolley,1937). The eclipse of May 29 1919 began in Brazil just south of the equator, with the totality track leaving the continent near the Amazon mouth, then swinging across the Atlantic to cross over the Portuguese island of Principe in the Bight of Benin, and then over tropical Africa in a long southward curving sweep. In accordance with what we have said above, the fifth eclipse in the series, the one on June 30 1973, also began in Brazil, but farther north, passed over the Azores and made land in the Saharan republic of Mauritania, continuing over Mali and Niger—these then being fairly newly independent territories of former French West Africa—and then over Kenya to end in the Indian Ocean. We can remark that, according to protocol, this last eclipse was one of a nineteen year series, all on June 30, which took place in 1935,1954, 1973, and 1992, all, incidentally, being visible from a well chosen common site. (Figure 8). We are particularly interested in the eclipses of 1919 and 1973.

# The Einstein Shift

It may seem incredible to a reader at the turn-over of the twentieth century that there was a time when the name of Albert Einstein was unknown except to a limited number of individuals of no particular consequence. He was a modest German civil servant in the Swiss Patent Office in 1905, when he published three papers on problems in physics. There was one on Brownian motions, the stochastic movements of very light-weight particles in a liquid medium, which demonstrated to the eye that particular molecules were bumping these test particles in random fashion. Then there was one on the absorption and emission of radiation by atoms, which did indeed attract attention and praise, and has been a permanent contribution to physics ever since. This is the one which, in due course, earned him a Nobel prize. (Unsigned, 1955).

Then there was one with the remarkable title, *Der Elektrodynamik Bewegter Körper*, (or The Electrodynamics of Moving Bodies), which indeed, embodied the concepts known as the Special Theory of Relativity. In spite of the fact that this paper offered new concepts of physics and new principles of measurement, it did not at once, get a warm welcome. For one thing the Dutch, Hendrik Antoon Lorentz, and the Irish, Gerald Francis Fitzgerald, had already speculated that such changes of physical parameters as those introduced by Einstein were necessary to account for the anomalous behavior of light waves as revealed in the famous Michelson-Morley experiment. This demonstrated that the velocity of light would be measured exactly the same by all observers, whether stationary or in motion, whatever these adjectives might actually mean.

In 1916, Einstein extended his analysis to include accelerated motion and the effects of gravitation. The time was not propitious for celebrity in the USA, even though communication with Germany was not seriously interrupted until their entry into the hostilities of the First World War. Surprisingly enough, the work of Einstein was better known in Britain through a personal connection of Einstein with the (neutral) Dutch citizen, Willem de Sitter, and his contact in Britain, at the Cambridge University Observatory, Arthur Stanley Eddington. (Figures 9 and 10).

Figure 9: Portrait sculpture of Albert Einstein
by Robert Berks, Washington D. C.
(The first equation on the scroll covers the
prediction of the gravitational deflection of light)
(Photo: Dr. G. W. Van Citters)

Figure 10: Arthur Stanley Eddington, (1882-1944),
Plumian Professor of Astronomy, and Director of
the Cambridge University Observatory.
(Royal Astron. Soc. photo.)

In spite of the formidable difficulties of communication of such an esoteric subject, Eddington felt that he had mastered it, and caused several papers by de Sitter to be published in the *Monthly Notices of the Royal Astronomical Society.* Eddington acquired the reputation as being the only man in England who knew what Einstein was getting at, and in due course he was asked to prepare a report on the General Theory of Relativity which was published by the Physical Society.

The theory replaced Newtonian gravitational attraction with the idea that massive bodies produced a change in the character of their surrounding space which caused other bodies to move in curved paths. The theory included three important predictions, one of which had already been verified. This was that the postulated space curvature round the Sun caused a sensible, though very small, systematic change in the orbit of the planet Mercury. In spite of the fact that the calculations of these orbital motions had been carried out by earnest gentlemen, using the primitive methods then available, that is, paper, pencil, and trigonometrical and logarithmic tables, in efforts lasting weeks, months or even years, there was general agreement that there was this anomaly, and even its value, and great efforts had been made to account for it, even to the extent ot postulating, and searching for an intra-Mercurial planet, Vulcan, which of course had not been found, because it did not exist.

The next prediction was that a spectrum line, or indeed any other optical phenomenon of a standard frequency, would, if generated in a strong gravitational field and observed from some external point, show a displacement of the frequency to lower values, i.e. a red shift. The existence of this phenomenon is now well-attested, especially in its extreme form of the black hole, but for long there was no sufficiently strong gravitational field situation offering the opportunity of observation.

The prediction of the greatest immediate interest was that light waves, perhaps more in their guise as photons, passing through a gravitational field would have their paths deflected. This idea was not entirely new : a calculation for an ideal photon, treated as a particle traveling at the speed of light, and passing close to the surface of the Sun, would have its path deflected, using pure old-fashioned Newtonian mechanics, by an angle of 0.87 seconds of arc. This looks pretty small, but by the standards of contemporary astrometry, and regarding this as a displacement of a star from its normal position, if it could be seen close to the solar disk, was easily measurable with precision. Einstein's theory using the idea of a distortion of space by the presence of the massive Sun, came out with a value twice as great, 1.75 arc seconds.

In the normal way of things, stars cannot be seen in the day-time close to the Sun, but this might be possible during a total solar eclipse, with the Sun temporarily hidden. Eddington was a highly esteemed, especially, but not entirely, theoretical astronomer who had been appointed to his prestigious double post of Observatory Director and Plumian professor of astronomy when in his early thirties. He was a bachelor, looked after in the official residence by his sister, physically fit, a former field hockey player, (as an amateur of course), and an habitual long distance cyclist. His only defect was his myopic eyesight. These features are noteworthy because towards the end of the First World War the slaughter of manpower had become so severe that even a man in Eddington's situation was bound to be conscripted into the armed forces. He, however, was a devout Quaker and had made it clear that if it came to the point, he would declare himself to be a conscientious objector, a status at that time often incurring severe hostility, with none of the more modern concept of alternative service. His distinguished colleagues, including his former superior at Greenwich Observatory, intervened, and he was given a deferment on the somewhat specious grounds, as it then seemed, that his scientific work was of importance for the war effort. What he was, in fact, planning, was the testing of Einstein's third prediction at the great eclipse of May 29 1919, in the fervent hope that the war might be ended sufficiently far ahead of that date to allow a restoration of civilian travel in the new political situation. It did, in fact, turn out that way. With the armistice of November 11 1918, Eddington's possible conscription fell away, and he was free to take a dominant part in the observation of that eclipse. (Douglas, 1956).

## The 1919 Eclipse Examined

We now take a much closer look at the circumstances surrounding that event. First, of course, the astronomical ones. One cannot choose where on the starry sky an eclipse will take place. The date determines the position of the Sun on its ecliptic track among the stars. The date of May 29 seemed particularly propitious since every year on that day the Sun moves against the star background through the outer parts of a cluster of quite easily visible naked eye stars, the Hyades. That cluster includes, quite by accident on the sky the bright red giant star, Aldebaran, which has nothing to do with the cluster and is much farther off than the other stars in the area. For a trial whether the presence of the eclipsed Sun distorts the nearby star field, requires the choice of an eclipse taking place in a promising field, and this case it looked very promising indeed. There are in the whole circuit of the ecliptic only six isolated bright stars which the Sun approaches closely on the sky in its annual motion. The Hyades region is a rare example of an ecliptic star field including moderately bright stars, of fourth and fifth magnitude and fainter. The totally eclipsed Sun is not perfectly dark: its bright disk is hidden, but is still surrounded by the solar corona, so it needs a distinctly bright star to be visible at eclipse fairly near the hidden Sun, and the relative brightnesses of the Hyades stars were in their favor. So this was the great opportunity:—an outstandingly long event at an especially favorable region of the sky. (Figure 11).

The politics of selecting an observing site, which might, in a war-torn world in other circumstances have been difficult, in this instance were favorable. The totality track started in north-eastern Brazil; in the war Brazil had been originally neutral, having a large immigrant population of German stock, but had later swung to the Allied side, and had not suffered any devastation as the result of so-doing. Perhaps the fact that the first Emperor after the end of Portuguese rule had been an amateur astronomer, might have caused the authorities to look favorably on a proposed expedition to their territory, even though the dynasty had been short-lived and replaced by a republic, or rather a confederation under the title of the United States of Brazil. Even so, the choice of the national flag showing the starry sky as seen from Rio de Janeiro at the moment of the establishment of the republican form of government may have had more to do with it. The totality track curved across the mid-Atlantic into the Bight of Benin, and just before touching the African continent would pass over the tiny Portuguese island of Principe, before continuing into the tropical rain-forest jungles of French Equatorial Africa and the Belgian Congo (as these territories then were). There would be no difficulty about going to Principe, geographically also a piece of remarkable good fortune, since a few miles farther north or south would have made it useless. Portugal, its political master, was sometimes referred to as England's oldest ally because of a long ago dynastic connection. More to the point perhaps were the economic links, epitomised by the proliferation of British, (or more correctly Scots), names in the lucrative port wine trade. Portugal had been on the Allied side in the war, but most of the consequent activity had been centered on her East African colonies. So, in due course, under the aegis of the Joint Permanent Eclipse Committee of British Astronomy, modest expeditions were sent to Sobral in Brazil and to the island of Principe.

## The Observing Expeditions

The war might be over, but certainly not its consequences. At Greenwich Observatory there was only one mechanic left after wholesale conscription, who doubled as a carpenter in the construction of a housing for the instruments to go to Sobral, the chosen site in Brazil, near Fortaleza. A large proportion of the almost innumerable small cargo ships, with occasional passengers, which, before the war, had plied the Atlantic Ocean, had been sunk by German unrestricted submarine warfare, but a ship, the Anselm, leaving Liverpool on March 8, was found to carry both expeditions to Madeira, and the Brazilian one on to that coast. Those destined for Principe had to wait for another local ship, and even the 'Brazilians' could not immediately get to their intended observing site. Those going to Brazil were Dr. A.C.D. Crommelin, a long-time member of the Greenwich staff with special expertise in the celestial mechanics of comets, but also a seasoned practical observer. He was accompanied by Charles Davidson, also a seasoned observer from the Greenwich staff. The local authorities gave the pair a warm welcome, and much practical help, and they were soon installed at the Sobral Jockey Club racecourse, with their equipment on level ground near the grandstand, which they found useful as offering covering for storage and assembly work. The pair were equipped with two lenses for the sky photography. One of these was the so-called Greenwich Astrographic lens, a doublet of 13-inches diameter, specifically designed for photography of the sky using the type of photographic emulsions then available. This was one of a large group of such lenses manufactured after the adoption some thirty years before of an international cooperative scheme to make precise astrographic charts by photography of the whole sky. Observatories undertook to observe a particular zone of declination on the sky and both Greenwich and the Oxford University Observatory had participated and completed their assignments, though this was not to be universally true of other sky sections for many years. The scale of the photographs made on plates 16 centimeters square, worked out at one minute of arc per millimeter, and since the measurements of star positions on such plates were made with machines giving a nominal accuracy of better than 1 micron (the thousandth part of a millimeter), it was clear that in ordinary circumstances the observers had the technical means to measure easily the kinds of star displacements envisaged by the eclipse program.

In normal use the lenses formed the objectives of permanently installed telescopes, equatorially mounted to permit pointing at any point in the sky, but re-installation of such heavy equipment in the field was never contemplated, and at Sobral, the astrographic lens and a back-up lens of four-inches diameter were installed in fixed positions on brick piers built for the occasion. The light from the eclipsed Sun and its surrounding field was to be fed into these temporary horizontal telescopes by means of devices called coelostats, which are flat mirrors mounted on rotation axes in their reflecting plane, and turning at half the rate of the diurnal motion, being powered usually by simple weight-driven clockwork. When suitably aligned the reflected beam from the sky field will pass in the fixed direction of the horizontal telescope. Two coelostats with 16-inch flat mirrors were used on the two expeditions. They had been made by Dr. A. A. Common, a pioneer of astronomical optics, donor of the Crossley 36-inch reflector to Lick Observatory and represented the acme of 19th century optical manufacture. (Dyson, 1904)

Both coelostats had been tuned up by E.T. Cottingham, a commercial clock manufacturer and pioneer of precision horology, who had long maintained close relations with the astronomical community. (Eddington, 1941). There was also an 8-inch one which went with the four-inch lens, both of which had been borrowed from the Royal Irish Academy. The axes on which the coelostats turned had to be pointed to the elevated celestial pole, and one of the coelostats had to undergo some re-education to adjust it to the southern hemisphere, Sobral being nearly four degrees south of the equator. The images presented to the telescopes were in fact inverted by the coelostats, but this, in itself is no great matter, since a common device of those engaged in astrometric studies was to produce an inverted image of a celestial field by photographing through the glass backing of the plate—with a small change of focus required—and then to compare this with, either the same or another field taken under different circumstances by clamping the pair of plates film-to-film, so that, in the case of repetition of an identical field, every star image turned up next door to its clone, and measurement could concern itself only with a study of the small positional shifts so revealed. In addition, the field of the image produced by the coelostat rotates, but this is no great matter if exposure times are short enough for the slight displacements to be ignored or coped with. Of more importance is the question of the true figure of the coelostat flat mirror, because deviations from ideal form are imposed on its image of the star-field. This did turn out to be a problem at Sobral with the 16-inch device, and recourse was had to stopping down the camera lens to 8-inches in an effort to improve definition. Such work is also vulnerable to wind shake or periodic error drive-rate of the coelostats.

At the eclipse, the weather was rather good, except for some temporary thin cloud, and the observers got 19 plates with the astrographic lens and 8 with the 4-inch lens. Everything was dismounted on June 7 but the observers returned on July 9 in order to obtain comparison plates of the eclipse field with the re-installed equipment, unfortunately at fairly low altitudes above the horizon—some 30 to 40 degrees—the Sun of course, now being absent. They reached Greenwich again on August 25.

Meanwhile, Arthur Eddington, accompanied by Cottingham, had been carried on to Principe from Madeira, where they too were met with a warm welcome and liberal practical assistance, being installed on a plantation, with some of the equipment having to be transported on the backs of local carriers through a wood. This installation was much like the one in Sobral with specially-built stone piers and a feed from the 16-inch coelostat to the mounting of an identical astrographic lens, this one borrowed from the Oxford University Observatory. The astrographic lens was stopped down to 8-inches to improve definition. All the remarks made above in connection with Sobral apply equally well here. However, in this case, the weather was not propitious. Eddington speaks of the *gravana* having arrived, by which was meant a dry wind with clouds, following the summer rainy season. We shall hear more of this later. The phenomena are associated with the quasi-weather front, which follows the seasonal movement of the Sun in low latitudes, and provides rain in the headwaters of the Nile, sandstorms in the Sahara, and so forth. It is usually called the Intertropical Convergence, and, being a meteorological phenomenon is not rigidly tied to the motion of the Sun. It had evidently passed Sobral and Principe, which had an unusual thunderstorm on eclipse morning, following a very cloudy period. In spite of drifting cloud, the observers carried out their intended program and obtained 16 plates of the eclipsed Sun, some showing a very fine prominence on the solar limb. The Sun was fairy active, with a strong corona, being just past a maximum (in mid-1917) in the eleven-year sunspot and activity cycle.

It had been intended to complete all the measurements of the plates on the spot, but the observers were compelled to leave by threat of a shipping strike and the risk of being marooned on the island for several months. After trans-shipping at Lisbon they got to Liverpool on July 14.

The problem faced by the enterprise was to compare the positions of stars with the eclipsed Sun present and the positions of the same stars when the Sun was absent. Before the Oxford astrographic lens was dismounted it was used to take a series of check plates, and another check field was photographed at Principe, but the observers could not wait out the many months until the eclipse field reappeared at night in the same sky position that it had occupied on the critical day.

In so far as it is impossible to obtain check plates of the relevant sky region at the same time with and without the eclipsed Sun, one must look at the differences and their possible effects on the measurements. Apparent star positions are affected by the time of year, by the altitude above the horizon of the sky region observed, and some other causes. One very important factor is a possible change of scale of the plate. Telescopes and other equipment are sensitive to ambient temperature and even to solar heating, which at an eclipse may be seriously reduced only just before the critical exposure time. Temperature changes cause an expansion or contraction of equipment which slightly change the scale on the photographic plate, and move all star images radially to or from the plate centre proportionately. In addition, in the case of a total solar eclipse, the light from the corona surrounds the eclipsed Sun, and a star image impinging on this and, so-to-speak having its light mixed in with the radially increasing coronal light, will have its image moved slightly towards the higher intensity.

When the 1919 plates came to be measured, the best turned out to be those obtained with the 4-inch lens at Sobral, and gave a deflection for a star at the limb of the Sun, inferred as 1.98 arc seconds. The Principe observations were interfered with by cloud and gave 1.61 arc seconds for the same quantity, while the Sobral astrographic observations gave 0.93 arc seconds.

It was decided not to attach much weight to this last result, and a combination of the other two values seemed to hit off Einstein's value of 1.75 arc seconds pretty closely. Eddington was firmly convinced. There was world wide publicity, which had the effect of promoting Einstein from a rather obscure, but respected figure, on to the world stage, especially in the USA, and fired up no end of public discussion, serious and jokey, of this new Relativity Theory.(Dyson, Eddington & Davidson, 1920). (Figure 12). Incidentally, the result had a profound effect on the thinking of Eddington, a deeply religious man, who came to believe that the mathematical apparatus of the Relativity Theory was the only one designed by the Creator for the description of the Universe. (Evans, 1998).

## Did Eddington Get it Right?

The critics were not so sure. The problems were expressed in a series of questions:

Were light rays passing near the Sun deflected?

If so, was the cause gravity, on the Newtonian model, or on the Einstein model, or refraction in a thin solar atmosphere?

If so, was the magnitude of the effect reduced in proportion to the radial distance of the light rays from the solar center?

If the magnitude of the effect at the solar limb matched the Einstein prediction, was this a proof of the correctness of the General Theory of Relativity?

S.A. Mitchell, in his book "Eclipses of the Sun" (Mitchell, 1923) covered all of these questions. He considered the results from two other eclipses, a well-observed one, by a party from Lick Observatory in Australia in 1922, and another in 1929. There were always the reservations that the star field of the 1919 eclipse, though exceptionally bright, consisted only of stars on one side of a line drawn through the eclipsed Sun, and the Lick results paid especial attention to the problem

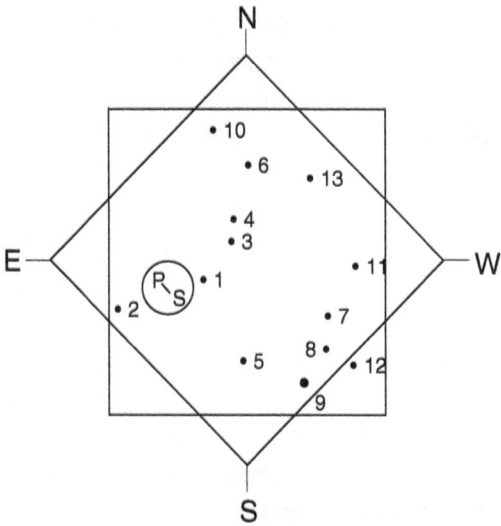

Figure 11: The star field for the 1919 eclipse. The Sun was at S for the Brazil observations, (oblong plates), and at P for the West African ones. (square plates). From Eddington, 1920)

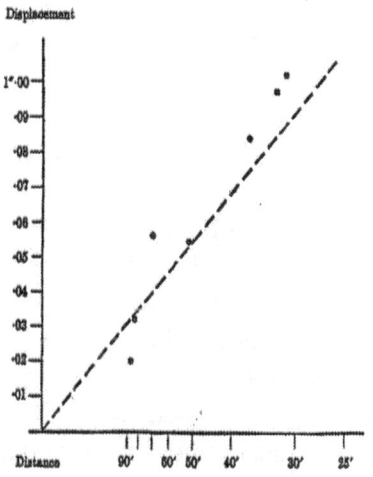

Figure 12: Final results for the 1919 expeditions, close to Einstein's prediction. (from Eddington, 1920)

of a radially asymmetrical starfield. Mitchell's result was the rather odd one, that he believed in the star displacements at their Einstein value, but he wasn't convinced of the correctness of the Theory of General Relativity.

A much more sophisticated analysis of the problem came from a contribution by H. von Klüber, (von Klüber, 1960). von Klüber reviews the results from no less than 12 eclipse determinations of the Einstein deflection and identifies what may be called the central problem of the whole business.

It can be supposed that one way or another, whether nearly the ideal one of check plates taken with the same instruments under the same physical conditions, or less directly, a system of star positions for the eclipse field has been established, to be compared with that taken at the eclipse. The scales of the two plates must be reconciled, and this leads to the determination of radial displacements of images, proportional to the distance, **r**, from the plate center. After these have been corrected one seeks the effects of the gravitational field of the eclipsed Sun, which it is assumed are inversely proportional to the radial distance, i.e. to 1 / **r**. So now we have star displacements which may be represented by some such formula as

$$\Delta = \alpha / r + \beta \, r$$

where $\alpha$ and $\beta$ are constants. Statisticians deal all the time with quantities which depend on two subsidiary quantities, as it might be the price of bread, and the price of butter, in the combination bread-and-butter, but the key point is that the two sets of basic data are usually independent. Here we are going to say that we will consider the shifts of stars a long way from the eclipsed Sun, where the first term is negligible, and the second shows up the scale correction. Having done that, and derived a corrected series of positional shifts we shall see how they compare with the hypothesis that they go like 1 / **r**, and from this derive the constant $\alpha$ which is the one which will tell us what the shift would be if this formula were valid right up to the limb of the Sun, where $r = 1$. But, hey, hey, you can't have it both ways and treat $r$ and $1 / r$ as if they were independent quantities, and you could choose which interpretation to put on them. What is more you are trying to find out whether the displacements really go like $1 / r$ so in a way you have already prejudged the results.

Let us swallow some of this and assume that somehow the scale has been corrected, so that we have a series of radial image displacements and the values of $r$ which go with them. von Klüber does this for all the data sets reported by the various eclipse expeditions.

We start with relatively large values of $r$, but not the very biggest, which we have already used up in determining scale errors, and reduced them to zero. Relatively large values of $r$ mean quite big circles on the sky and if we are lucky with the star distribution in the eclipse field, there will be lots of stars available for determinations of the relation between distance and displacement. They should show a gradual increase with decreasing distance from the Sun, and if one didn't know or assume better, a straight line fit would probably do very well. As we move closer inwards the area of the star-field shrinks, there are fewer and fewer stars in the field and the curve become less and less well-defined. The relation between displacement and radius is, in fact, hyperbolic and soon, somewhere, this curve has to take a sharp turn upward and increase rapidly towards the critical value at $r = 1$. Of course, in our application, the curve can stop there but geometrically the hyperbola shoots up to infinity for values of $r$ less than unity. When we get to this critical turn the area of sky concerned is so small that there may be very few or no stars in it, and tracking where the curve goes becomes a matter of extrapolation, since the data we would like to have, have quit. This means that in most cases the final answer depends critically on perhaps one or two star images relatively close to the Sun, where the effects of exposure to the light of the corona are most pronounced

(Figures 13 and 14). When this procedure for verifying Einstein's prediction was first proposed it sounded not too difficult but it has turned out to be a tantalizing problem.

Among his rather sad conclusions von Klüber says, "All observations clearly show that a light deflection of the kind expected quite obviously exists in the neighborhood of the Sun. But the observations are not sufficient to show decisively whether the deflection really follows the hyperbolic law predicted by the General Theory of Relativity, mainly because, so far, it has not been possible to obtain a satisfactory number of star-images sufficiently close to the Sun. As things are at present, most observations could be represented quite well even by straight lines." After listing other desiderata at a well-run eclipse, he concludes with the words "Further observations are only justified if real progress is to be expected as result of fulfilling as nearly as possible the stringent conditions summarized above. Even so, such observations remain among the most difficult of all those which can be attempted at a total solar eclipse."

## The Challenge: Dramatis Personae

The eclipse of June 30 1973 naturally came to the attention of many astronomers, for perhaps a majority of them, of no more interest than many another, but for a minority, perhaps with a sense of history, there was a possibility of repeating Eddington's observations on an eclipse having the same outstanding quality, and even with something like a repetition of its physical circumstances. But if anything were to be done, it would be in circumstances quite unlike those which obtained in Eddington's time.

The space age and the atomic age had arrived. There were many more university faculties of astronomy, astrophysics and physics than there ever had been in Eddington's day, and many

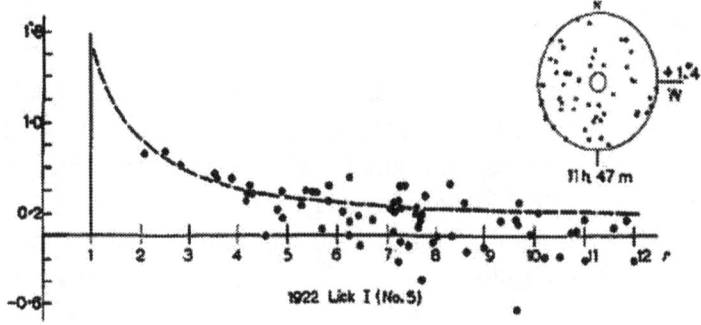

Figure 13: A relatively good determination of the
gravitational deflection made at an eclipse in 1922.
(from von Klüber, 1960)

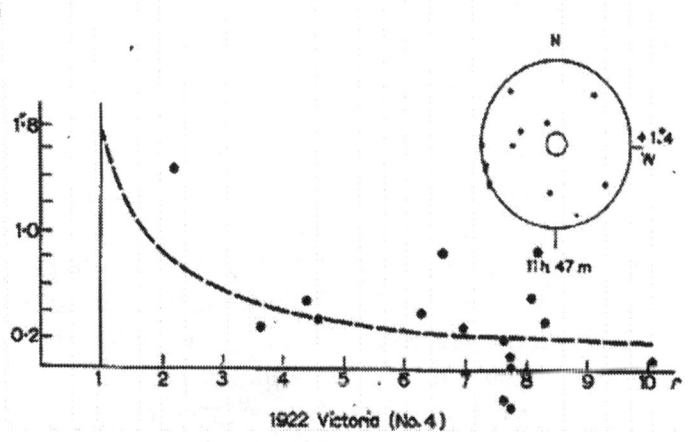

Figure 14: A less reliable result from another 1922
eclipse, where the displacement curve is ill-
determined close to the Sun.
(from von Klüber, 1960)

more astronomers. In 1938 the International Astronomical Union had held its General Assembly in Stockholm and the attendance was about 400. Now an annual meeting of the American Astronomical Society alone has to cope with almost ten times that number. In addition, amateur astronomers had become far more sophisticated and numerous, no longer restricted to optics they could manufacture themselves, but equipped with high-quality instruments purchased from specialist companies founded to meet the growing demand of a relatively affluent class. The shipping companies had latched on to the possibilities of sea-borne observations for amateurs, and, with selected professional tour guides, could promise, as well, the chance of a view of totality, enhanced by detailed meteorological forecasts and the possibility of sailing the ship to an optimum location. Air travel had become commonplace and people could now get themselves transported to some out-of-the-way places, and even carried farther in terrain-worthy motor vehicles such as the British manufactured Land Rovers.

These considerations were to become important, especially in view of the totality track of this particular eclipse, which would enter Africa and traverse the old colonial territories of the French empire—the *pays francophones*—which were now independent, though in many cases retaining links with what had been the mother country, in others subject to tribal and political rivalries even making travel unsafe.

Texas was not, as we shall see, by any means the only institution interested in the eclipse, though it turned out to be the only one with a program of observation of the Einstein deflection. How this came about is one of those stories of an almost accidental meeting of minds and resources, originating from a variety of sources.

As is often the case, the story starts in the fairly distant past. William Johnson McDonald belonged to a settler family that immigrated to northeast Texas in 1837. A confirmed, and, eventually, somewhat crusty bachelor, after the Civil War he became a prosperous banker. Not perhaps notably an advanced thinker, he was undoubtedly in touch with some current scientific advances, since his library included books by such as Darwin and Huxley. He may have had some interest in astronomy, though one suspects that this was emphasized in the light of subsequent developments. He was certainly not an academic, and stories were to surface that he had a chip on his shoulder about what he thought were people who did no productive work. At all events, his lifestyle and opinions came under the closest scrutiny when his will was proved after his death at the age of 82 in 1926. After some modest bequests to members of his family, the whole of the rest of his estate, amounting to something like a million dollars was bequeathed to the University of Texas for the foundation of an astronomical observatory. The will was, of course, contested by the family members, on the grounds of insanity: It was stated that McDonald had said that if one had a sufficiently large telescope one could see into Heaven and see who was there. On the other hand he was cited as a prudent and successful banker, and an amateur botanist who had even in mid-life taken some summer courses at Harvard. The cases dragged on, with the University intervening to protect their interest, and finally in 1929 they prevailed and received $ 840,000 when the various actions were settled out of court.

Now came the problem that the University had almost no idea what to do with the money. The state constitution referred to the necessity for the foundation of a "university of the first class" but, at the time, only the most partisan supporter might have so described the institution. In fact, it might well be said that this bequest was an important first step in raising the university to the degree of eminence that it now enjoys. At the time there was only one faculty member with astronomical qualifications, E.J. Prouse who taught mathematics, there being no astronomy faculty, and no teaching of the subject much beyond the late Victorian positional style often current in small institutions. The situation was saved from what might have been a serious misuse of the funds, by the intervention of Otto Struve, a distinguished astronomer at the University of Chicago, who had become dissatisfied with the climatic conditions at his own, nearby, Yerkes Observatory, and was even considering whether he could establish an observing station in the clear climate of West Texas, or some other likely place. The enterprising Struve eventually secured what was then a most unusual compact between the prestigious private university of Chicago and the relatively obscure state university at Austin, Texas. (Evans & Mulholland, 1986)

Under this compact Struve was to supervise the selection of a site for a large telescope, its manufacture and installation, and operate it for 30 years by Chicago astronomers, while Texas could get its act together and create some indigenous astronomy. It all worked out remarkably well with some bumps along the way and, by 1963, Texas was ready to take over the telescope, with a fledgling department of astronomy at the university in Austin. The first Texas director and chairman of the Astronomy Department was Harlan J. Smith. Born in 1924, Smith lived for the first 18 years of his life on Wheeling Island, the only inhabited one on the Ohio River in West Virginia. Graduating from High School in 1942, Smith was runner-up in the first Westinghouse Natural Science Talent Search. In 1943 he enlisted in the US Army Air Corps as a meteorologist, a duty which took him to Harvard, where he renewed acquaintance with Harlow Shapley, one of the judges in the Westinghouse competition. Smith completed his Bachelor's degree in 1947 and embarked on a study of those short period variable stars which resemble Cepheids, now known as Delta Scuti stars. In 1953, after his marriage to Joan Swift Greene, daughter of a missionary, who had been born and raised in China, Smith was appointed Instructor, (later Assistant and Associate Professor), in astronomy at Yale

University. In due course he became a co-editor of the *Astronomical Journal* and acting secretary of the American Astronomical Society, but these responsibilities did not stop him from further researches on the Delta Scuti stars, a keen interest in radio-astronomy, and an important discovery of the optical variability of quasi-stellar objects. (Douglas,1992).

Smith's post at Texas called for a great deal of imagination and vision, with plans for updating the original 82-inch telescope, (Figure 15), the construction of a new one of 107-inches in diameter and a development in the field of radio astronomy. At Austin there were by now already four faculty members, and four support staff and a few graduate students. Smith was a great inspiration and recruited some important associates, including instrumentalist, Bob Tull, electronics specialist from industry, Ed Nather, and, at first on an NSF Fellowship, later as a faculty member, David Evans from the Royal Observatory at the Cape of Good Hope, who improved the administration of the observatory. A significant development was the organization of no less than three two-man field teams to observe a rare event in 1971, the occultation of a bright star by Jupiter, to improve knowledge of the highest atmospheric layers of that planet. This event could be observed from any place on Earth where Jupiter would be above the horizon, the ideal position, with the planet in the zenith, being in the middle of the Indian Ocean. So teams went to Australia, India and South Africa, with successful results, though one team was half clouded out and another totally. The importance of the three teams was that they were in unrelated meteorological areas. When it came to the 1973 solar eclipse, thinking was, at first, greatly influenced by the idea of multiple expeditions, with no realization of the importance of the disparate climatic conditions at the different sites in 1971.

So Smith felt he had an organization which could mount a successful field expedition for 1973, and he fortunately fell in

Figure 15: The legacy of William Johnson
McDonald. The dome of the 82-inch telescope on
a clear snowy night. (Photo Lee Anne Willson).

Figure 16: The stars of the show. Bryce and Cécile DeWitt still vigorous and intellectually active in the 21st Century

with two very important people in another University department. Alfred Schild was an Austrian Jew, born in Turkey, who, as a young man, fled Nazi Austria in 1939, and was interned, first in England and then in Canada. He was one of a group of similar individuals who used their unfortunate situation for the study of physics and astrophysics. In 1957 he came to Texas, where he created a world-famous school of relativity, statistical mechanics and particle physics, though his liberal ideas and a certain impatience with authority sometimes incurred grumblings from high quarters. This was the scene that, in 1972, attracted Bryce DeWitt, a Californian by birth, with degrees from Harvard, membership of the Princeton Institute of Advanced study, a Fulbright Fellowship and many other distinctions, later to be even more enhanced. In the latter part of World War II he had trained as a Navy flier. It was he who was one of the principal advocates of the proposal to observe the Einstein deflection at the 1973 eclipse, and would take a leading part in its planning and execution. (Figure 16)

In some ways even more important than Bryce, was his wife Cécile DeWitt-Morette, a French blue-stocking, whom he married after they met at Princeton. She had achieved her education while France was under German occupation, and endured the destruction of the family home at Caen during the D-Day landings, while she herself was in Paris. This charming lady, now a Chevalier of the Légion d'Honneur, had a brilliant career in the higher echelons of theoretical physics, and had founded, while still very young, a celebrated summer school of theoretical physics at Les Houches in the French Alps. Eventually mother of four children, she still enjoyed outdoor recreations such as kayaking. At the time of the eclipse Mme DeWitt was still a French citizen, with many contacts in the scientific administration of France, and contacts from school and university days both in France and in the former French colonies. She had come to Texas, at first associated with the astronomy department, because of so-called nepotism laws in Texas, which prevented her, for a time, but not forever, from holding a professorship in the same department as her husband. As we shall see, her contacts in the academic and political world of France were to be crucial in smoothing the way of the enterprise.

Two other early important recruits to the team were Alfred H. Mikesell and F. Burton Jones. Mikesell, an older man, had been first employed in the then meagerly rewarded astronomical world of the nineteen-thirties. He had been deviated from a purely academic career, by the discovery of his extraordinary talents as a highly qualified instrumental technician, and a photographic and photometric specialist. His vita reads like an astronomical adventure story, including an episode in a high-altitude helium balloon, but also, especially at the US Naval Observatory, several periods of major instrumental improvement, carried out without interruption of observing programs. His early association with Lick Observatory familiarized him with previous eclipse expeditions to observe the Einstein deflection, and the crucial importance of mastering the exact imaging on photographic plates to be used at an eclipse. He was, certainly, the most highly qualified man available in the arcane specialties needed for this task, and Texas was lucky to persuade him away from another project that offered. His record of meticulous hard work as coordinator of the expedition cannot be overemphasized.

F. Burton Jones was a young American, who, in the period before the eclipse, was a visiting scientist at the Royal Greenwich Observatory at Herstmonceux in southern England, and, somewhat typically of the administrative situation at the time, could be spared for the eclipse work, as long as they didn't have to fund him specially. Burton Jones was a newly-minted astrometric specialist, whose connection with Herstmonceux was important because of the development there of automatic plate-measuring machines which offered better accuracy and more comprehensive computer-driven reduction programs than had been previously possible.

# Basic Geography

If the eclipse were to be observed it had to be from somewhere. A close study of large-scale maps of the totality track was discouraging. The center of the totality belt would cross the African coast north of Nouakchott, capital of the Republic of Mauritania, and pass over the small towns of Akjoujt, Atar and the oasis of Chinguetti, after which it would head almost due east across empty deserts. Then into the Republics of Mali and Niger, where it would pass over the Aïr mountains, before swinging south-east over Chad, the southern part of the Sudan, and then over Lake Rudolf in Kenya, before passing out to sea and ending over the Indian Ocean. Except possibly the Kenya site, where the Sun would be rather low and the totality phase shorter, the territories were all more or less severely inhospitable and inaccessible, with almost no logistical support possibilities anywhere. Added to this, these territories which, up to 1958, had belonged to the French union, before gaining complete independence in 1960, were, in some cases, in a state of political turmoil, with opposition groups liable to try coups d'état. (Figure 17).

Whatever else might happen, clearly Mauritania would be of the greatest interest, presenting a variety of problems of different kinds. With a nominal area of 419,000 square miles and a population of 1,480,000, it had become fully independent in 1960, severing a previous link with the French Community, and replacing, in 1972 !on eclipse day !, the former currency common to the previous French West African colonies, the CFA, with a new unit, the ougiya. The president and prime minister, re-elected several times, was Moktar Ould Daddah.

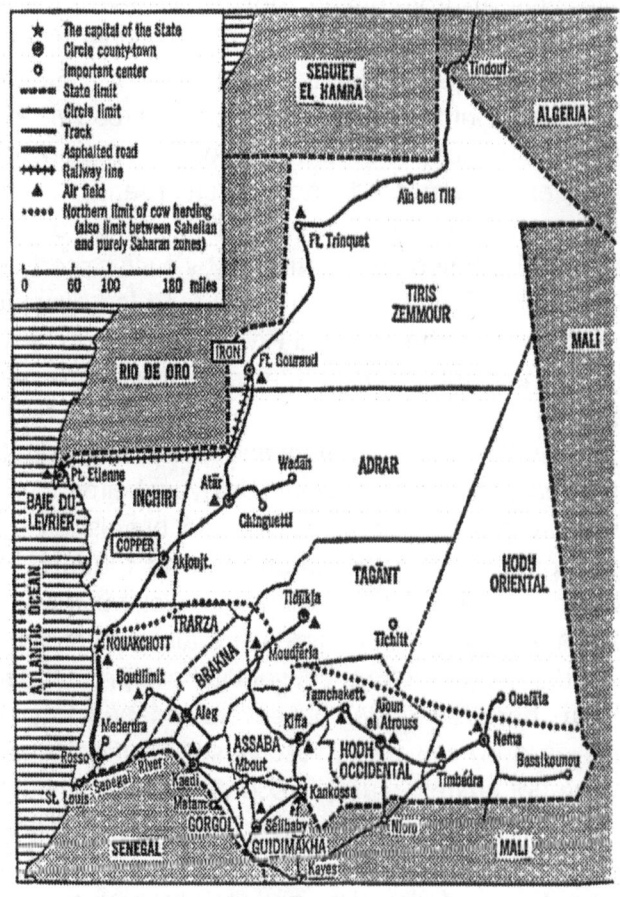

Figure 17: Sketch of Mauritania,
from Gerteiny, 1967.

These statistics do little to explain the real problems presented to the unfortunate Mauritanians by the prospect of a total solar eclipse on their territory. The country was divided into three separate and mutually almost inaccessible regions. To the north, via the port, Nouadhibou, (formerly Porte Etienne), was an iron-mining region with even a rail-line extending some way into the interior. At the middle of the coast lay the capital Nouakchott, with a population of 135,000 from which extended into the interior the important road via Akjoujt, and Atar to Chinguetti. Between the northern region and this region lay an uninterrupted roadless sea of sand, traversable it would seem on foot by a man leading a string of camels, but not always, based on experience, by even desert-worthy motor vehicles. To the south, a coast road led down to the Senegal border at the Senegal River, to the north of which, mainly to the east lay a number of scattered oases and settlements.

The Mauritanian government had reason to fear that their exiguous natural resources would be overwhelmed, not only by large foreign expeditions of scientists, but also by groups of astronomically-minded amateurs and tourists. Some contacts had been established at high level, but it was clear already in early 1972, that there were all kinds of agencies interested, not only the French, said to have pre-empted the old military barracks at Atar, but several independent foreign universities, and even a group of diverse US official agencies, who were not entirely mutually coordinated. In other words there seemed to be a superfluity of cooks in this particular broth.

The defining meeting of the professional astronomers had occurred on May 15 in Madrid, under the chairmanship of Cornelis de Jager of the Netherlands, who was coordinator of the eclipse for the International Astronomical Union. It was attended by two government representatives:

Dr. El Sammani A. Jacoub, Eclipse Coordinator, National Council of Research, P.O. Box 2404, Khartoum, Sudan.

Mrs. Cheikh Abdallahi, Département d'Artisme et Tourisme, B.P. 246, Nouakchott, Mauritania, West Africa.

Mrs. Abdallahi represented the Secretary-General of her department, Mr. Ahmed Ould Die, who was responsible for eclipse arrangements in his country. The other governments of moment were unfortunately not represented.

The attendees submitted the following plans of intended activity, showing in each case the number of persons to be involved. All were ground-based except as noted.

| | | |
|---|---|---|
| Houtgast | (Netherlands) | Mauritania (5) |
| Menzel | (USA) | Mauritania (50) |
| Vila | (France) | C.Verde Is and Ascension (4) |
| Weldmeier | (Switzerland) | Mauritania |
| Hagen | (USA) | Mauritania, Mali or Niger |
| Arnquist | (USA/Italy) | Airborne or Mauritania (5) |
| Sykora | (CSSR) | Niger (7) |
| Korff | (USA) | Mauritania (2), Kenya (2) |
| Steshenko | (USSR) | Mauritania (15) |
| Rösch | (France) | Mauritania (12), Niger (3) |
| Rigutti | (Italy) | Niger (5) |
| La Count | (USA) | |
| | NASA Optical | Niger (14) |
| | ONR | Niger or Kenya (6) |
| | 7 universities, | Mauritania and Kenya |
| | NOAA | Sudan or Chad |
| | NSF (up to18 universities) 2 sites, (100) | |
| | NSF airborne | Mauritania, Kenya or Niger |
| | NSF rocket | Mauritania |
| Rycroft | (England) | Cape Verde Is, Mauritania (5) |
| Hiei | (Japan) | Mauritania (12) |

It was doubted whether there were any serious groups unrepresented in this list, from which it was clear that Mauritania would be bearing the brunt of the assault of those wishing to observe the eclipse. Mrs Abdallahi outlined some of the measures which a committee established by the Mauritanian Department of Industrial Development was proposing to take. These included special customs facilities. It was proposed to restrict the number of entrance permits because of the limitations of local facilities, to insist that all parties remain near the three towns cited, unless they demonstrated willingness and ability to take care of their own needs. Accommodation would in most cases have to be in tents. Amateurs were going to be a serious problem—they had received over 1800 enquiries from France alone. Some areas might have to be closed to amateurs.

Rösch raised the question of contamination of the sky from aircraft, prompting a claim from La Count of six hours as a desired dissipation time from contrails. There were no internal scheduled flights on a Saturday according to Mrs. Adallahi, but this might not apply to international flights overflying Mauritania.

Dr. Jacoub, in view of the lack of interest in the Sudan situation spoke briefly of such matters as customs waivers, and said that the prime site in his country would be Hofrat en Nahas, a newly developed copper mining area right on the center line, accessible by car from Khartoum in 12 hours. Although there was a fairly long section of eclipse track within the Sudan, close to the southern border, the points might have been raised that this was close to the sudd area of marshes, lagoons and forests in the headwaters of the White Nile, and that a copper mine might easily be a contaminator of the sky, but the discussion was sufficiently encouraging that one or more expeditions might reconsider their decisions for Mauritania.

G.W. Curtis (NCAR-NSF) reported preliminary site surveys in Mauritania, Niger and Mali. In every case the problems were the same : sand, heat, supplies, lodging, transport, poor roads. All unloading would have to be by hand, so that careful planning and staging would be imperative. Sites would be revisited in June, and stops would also be made in

Kenya, Chad and the Sudan. The NSF site or sites would be decided in early August.

This report came from Derral Mulholland, from Texas, who could not attend the second session because of a commitment to his own working group of the IAU, quite unrelated to the eclipse.

# The DeWitt Odyssey

For most of the aspirant observers, their designated eclipse sites represented no more than hopes for unknown points on maps, even though, as has been mentioned, there had been a certain amount of investigation in the field. In the Texas case, Bryce and Cécile DeWitt set out to investigate possible sites on the ground, in the first instance with the implied plan for three separate observing sites, together with suggestions by Wilkinson and others at Princeton, who were considering cooperation in a Texas enterprise, who thought that the mountains of the Aïr massif in Niger looked particularly promising.

Bryce kept a detailed diary of this enterprise, some parts of it surprisingly frank and little related to the task in hand. After the journey was completed he sent a copy of his hand-written text of over a hundred note-book pages, to Al Mikesell for the consideration of the committee in Austin. Mikesell removed some of the material he judged less relevant to the question of site selection and submitted this Bowdlerised version to the committee. The present writers feel that the unabridged version gives a much more interesting picture of what went on, and quote extensively from it, with only a minimum of censorship. It must be made clear that at the beginning Texas thinking was for an independent expedition or expeditions and that, only later, did this enterprise move into the scheme of general logistic support provided for American expeditions. Thus the DeWitt survey was taken independently and financed by a grant from the Research Corporation to the Department of Astronomy.

The diary begins on Saturday May 27 1972, with Bryce succeeding in talking by phone to Bill Curtis (of the National Center for Atmospheric Research) in Dakar, Senegal. The latter had been in Mauritania, but his Land Rrover had broken down leaving him stranded for two days. His opinion of Chinguetti was unfavorable, saying that the road to it was very bad, that it had nothing to recommend it to the attention of the NSF as an observing site, in spite of the statements that Chinguetti was the seat of an ancient Arabic university and library, of which we shall hear more. The words "university", "library" conjure up particular sorts of image in the western mind, whereas the reality was the existence of a collection of manuscripts—some known to modern scholarship—held in a tiny mosque in an oasis with a population of, one would guess, no more than 3000 people. The opinions about roads also were found to reflect directly the degree of inexperience with outback country travel in Africa. The opinion of Evans, formed later, was that the road wasn't too bad, with maybe some drifting sand patches, and a bad stretch on the steep Amogjar Pass, but even that could be negotiated by local trucks.

The primary reason for Bryce's call was to try to secure a place on a proposed expedition to the Republic of Niger, with especially, an investigation into the suitability of a site in the Aïr mountains, since both Harlan Smith and Dave Wilkinson thought that a higher altitude site would be advantageous. Curtis emphasized the difficulties, but said that if Bryce and Cécile could find a vehicle in Agadès, an important town in Niger, he would be happy to have them join the caravan. This put the idea in Bryce's head "how we could make use of our tenuous contacts in that part of the world", and that they would have to "scrounge like hell" for themselves.

On May 31 he presented a copy of the Texas proposal to Dr. Harold Lane at the National Science Foundation—"who was rather stunned by the proposed budget" which was hovering in the region of $ 400,000. There also seemed to be a good deal of confusion as to whose baby relativity was, though there was great interest in the forthcoming African expedition of the DeWitts.

He was to see Dr. Ronald La Count, who had been principal USA representative at the Madrid meeting, in the afternoon, after a luncheon with the Mauritanian Ambassador. First at the Embassy they obtained Mauritanian visas and then were taken by black Cadillac to the ambassadorial residence. Mr. Moulage el Hassen, of nomadic ancestry, accompanied by his wife and family, and a Mauritanian medical guest, entertained them in traditional style seated on floor cushions, to a typical meal of mutton with semolina and hot sauce. Wine was offered to the non-Islamic guests. After a fruit dessert they repaired to the salon where they made first acquaintance with the obligatory ceremonial tea-serving, of which more later, and began a useful discussion of conditions to be expected in the desert.

These were mainly concerned with climate, which, in spite of a reference to a rainy season by the ambassador, seemed to include the unspoken assumption on the part of the visitors, that bad weather in the Sahara was sporadic and spotty, so that occupation of more than one observing site might provide an advantage. What the ambassador had to say about sandstorms seemed to contradict an opinion by Mrs. Ould Daddah, the highly placed Mauritanian lady, whom Cécile had encountered at a meeting. All the major places were accessible by domestic airline. Electric power at Chinguetti was available, though only to the government—again a discrepancy between Curtis and Mrs. Ould Daddah. Dr. Touré, the medical visitor, recommended getting in touch with a Mauritanian astrophysicist, Dr. Alassane Sy, a rare enough species, now resident in France. The visitors were duly returned to the Embassy, with a parting remark from the Ambassador that he would soon be in Mauritania and ready to offer help, as well as giving the names of the Secretary-General of Tourism, in charge of coordinating eclipse activities and the head of the Ecole Normale Supérieur in Nouakchott. This incident illustrates the particular advantages enjoyed by the De Witts through their many contacts, which, perhaps, made things smoother for them than for some of the other individuals involved.

Later in the afternoon with Dr. Ronald La Count at the National Science Foundation, they had a most useful meeting, though Cécile had to leave early to catch a plane. La Count was most anxious for the NSF to be as helpful as possible, especially in smoothing relations with foreign scientific teams, and host countries. Bryce asked him to phone the French scientific attaché in Washington to acquaint him with the offer by Cécile to teach at Nouakchott, or elsewhere, as a goodwill gesture to host nations.

He was, at first, somewhat reluctant, until it was made clear that this did not commit the NSF to anything. Although supposed to be in charge of site survey and logistics, he was unhappy at being by-passed. The DeWitts told him that they had been informed that Bill Curtis was the man concerned. There seemed to be a possible rivalry between Curtis and La Count or between NSF and NCAR. It was not clear who had first mentioned Chinguetti. As far as the DeWitts were concerned it was in a conversation between Cécile and Mrs. Ould Daddah, but La Count's response was that this was the first place that everybody thought of. It was during this exchange that Mrs. Ould Daddah had stated that it would simply be impossible for Mauritania to handle large numbers of people. Bryce was dumbfounded to discover that quite apart from anything he himself had done, La Count was arranging for him to visit the Aïr mountains in Niger, along with Douglas Buritt of the Nasa-Goddard Space Flight Center, who would be driving a Land Rover from Niamey, the capital, to Agadès about June 19, to determine the condition of the road. At that moment Burritt was in Canada and unavailable on the phone. La Count had been in Niger the previous year, when the dollar was devalued, and had had much trouble in arranging necessary funds.

He was interested in a project for the preparation of a brochure in French and Arabic, giving an account of the history of eclipses and especially of the Arabic contributions to astronomy, the latter in collaboration with acknowledged Arab scholars to avoid all risks of giving offense. There had recently been a total eclipse of the Moon which had disturbed many of the desert nomads. As far as total solar eclipses were concerned, there would be a careful warning of the danger of incurring an eclipse burn, a subject already discussed with the Mauritanian ambassador who was enthusiastic, not least because he himself had suffered some damage to his eyesight from this cause.

If one tries to take a sustained look at the ordinary full Sun, the intense light immediately shuts down the pupil of the eye to protect it by minimising the energy incident on it, and causes pain which immediately commands the owner to look somewhere else before any damage can result. At a total or near-total solar eclipse, when perhaps only a tiny sliver of the solar surface can be seen, there is no overwhelming flood of radiation to cause the iris to close down and the gaze to be averted, but the small area of solar surface is tremendously bright and the eye acts like a burning glass, focusing an intensely hot image on the retina, burning a hole in it, and leaving the owner with a permanent blind spot.

On June 1, Bryce had gone to Detroit but recorded that La Count had telephoned him expressing his hope that visiting expeditions should do something for the local inhabitants, such as leaving behind as much as possible of the non-scientific equipment, but he emphasized that Bryce should not speak for the NSF on such matters.

While there, Bryce was in touch with Cécile who had been lecturing at Pennsylvania State University and had come in contact with John Hagen of the astronomy department, which was to send, as stated at the Madrid meeting, a large group, five or six, with a considerable amount of equipment loaded into a van. He was not yet sure whether he would be going to Niger or Mauritania. He discussed the relative freight capacities of the airports at Agadès and Nouakchott, and spoke of hiring a tractor in Dakar to haul his stuff to Nouakchott. He too, was considering leaving behind non-essential equipment after the eclipse—"on the QT" as Bryce puts it. He was anxious to have someone explore the possibilities of the Bagyan or Bagueyane, mountains in the Aïr massif, and offered $ 300 to help with Bryce's expenses.

By the first week in June, the DeWitts were in Paris in contact with French astronomers proposing eclipse expeditions. These included Rösch, Schatzmann and Labeyrie from Meudon, and a geologist from Orléans. Rörsch repeated to Cécile what he had already said in Madrid, namely that there would be 12 French scientists at Atar, and three or four more at the uranium mine at Madouala in Niger. He was most friendly but pointed out that the French had made these arrangements more than a year before, whereas the Americans had yet to choose their sites. He made a special point of this because there seemed to have been some bad feeling with La Count over the French pre-emption of the old barracks at Atar. He was afraid that the sheer size of the American effort might change the balance of the logistics for small expeditions. Rösch was in touch officially with the government department concerned with intellectual contacts with the former French colonies, which was clearly unlikely to be very helpful to an American expedition.

So the DeWitts went to work to exploit their quite extraordinarily influential connections, largely of course through Cécile. A French physicist gave her the name of Mademoiselle Adda, a French mathematician who had taught at the University of Niamey in Niger, who in turn, gave them the name of Professor Trichet, the geologist from Orléans. He proved to be an expert on the geology of Niger, including especially the Aïr massif. He depressed Bryce by saying, after a long discussion about the lack of maps, that it was difficult to get high in the mountains because of the extreme steepness of the peaks, but he did mention a village far into the region, El Meki. The DeWitts had supper with an old friend of Cécile's from Girl Scout days during World War II, a collaborator in clandestine activities during the occupation. She, a gynecologist had assisted at the birth of the first child, a son, of the wife of the President of Mauritania. This lady was French and had a law degree, but was not to be confused with "our principal contact in Mauritania", Mrs. Ould Daddah, the President's sister-in-law, a brilliant woman of Algerian origin. Cécile had met her in New York. She, apparently had difficulty in adapting to the conditions of life in Mauritania and suffered occasional bouts of depression, which may account for a rather cool response to a telephone call made to her by Cécile from Austin. Finally, by sheer accident, Cécile's brother, Jaques Morette, Chief of the Mission of Cooperation at Abidjan on the Ivory Coast, suddenly appeared and smoothed their way through some of the bureaucratic and diplomatic jungles at the Ministry of Foreign Affairs.

Bryce had some concern over bookings of tickets for journeys out of Mauritania after the eclipse, this being the season when everybody who could, headed for Europe. After accepting some delays, matters were arranged. They soon began taking anti-malarial pills, and equipping themselves with medicines for stomach upsets, and getting second shots for cholera. The first had been given in Austin, along with ones against yellow fever, typhoid and typhus. Bryce equipped himself with 5 rolls of 35 mm Kodachrome for use with the Questar telescope which he had with him. He found that he could run its drive with four large six-volt dry batteries connected in series. What with one thing and another, including a good supply of dehydrated food, their baggage was well over weight, but this was fortunately not insisted upon.

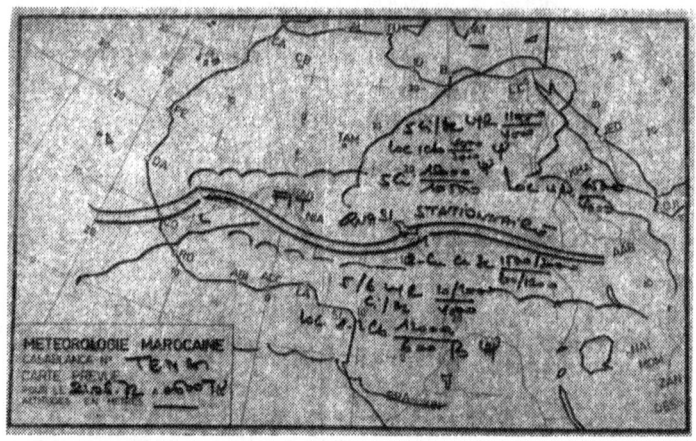

Figure 18: The Intertropical Convergence on the
meteorological map of West African for May 1972.
(Courtesy of the PanAm pilot)

Figure 19: How it looked from above.

It was at this point that they received from Mikesell the report of David Evans who, at the end of May happened to be flying down from Morocco, first to Lagos, and then on to Johannesburg aboard a PanAm plane. In striking contrast to modern conditions, Evans had sat in some empty seats and talked to the pilot, to whom he had sent a note.

Today, he would probably have been immediately overpowered as a highjacker, but the pilot, a surprisingly unimpressive gentleman with granny glasses, was most helpful when Evans pointed out that they were about to cross the eclipse path, not so far from expected ground observation sites, just about a month ahead of the important date in the following year. The pilot produced, and let Evans keep, his meteorological chart from Morocco, which showed a large disturbed area more or less along the eclipse track. This, he said, was the Intertropical Convergence, essentially a weather front which moved with the Sun and would almost certainly be on top of the eclipse track at the critical time. This, of course, ruled out any advantage of observing from several places in hopes that conditions would be uncoordinated. Evans also took a series of color shots out of the window showing the cloud band, which was some 500 miles broad, broken occasionally by rifts full of red dust. (Figures 18, 19)

Bryce was unimpressed by this news and noted that, at that date, Mauritania should be clear, and cited other examples of travelers who had found fine weather where, if the incidence of the Convergence was quite regular there would be clouds. He remarks, "There are certain general tendencies, (e.g. most days are sunny), but also wide departures from the norm".

Next day, they had lunch with Mademoiselle Adda, who brought letters of introduction to important people in Niger, and some gifts which she asked them to carry. Considering that they already had gifts for Mrs. Ould Daddah's children, and kaleidoscopes, produced by Marjorie Mikesell, Al's wife, to be used to maintain friendly relations with any children who might be wanting to look through the Questar, their baggage was even more augmented.

In the afternoon, Bryce telephoned Dr. Alassane Sy, whose connection with expedition was not yet clear, to find out whether he planned to be in Mauritania for the eclipse. He thought not, but suggested that they look up his uncle, Dr. Ba Bocar Alpha, a cousin of Dr. Touré, in Nouakchott.

In the evening they went out to Saint Germain-en-Laye to the home of an aunt of Cécile's, where they spent the night. Jaques Morette arrived next day for lunch. At this point Bryce describes Cécile's visit to the Secretariat d'Etat, where she had long talks with the Chef de Service for Mauritania and the Chef de Service for higher education and cultural exchanges. Cécile had explored such possibilities as getting more of the local people involved, but Bryce remarks "the Secretariat d'Etat does not envisage its mission as one of helping the former French colonies to strengthen their links with the United States.", and expressed the thought that there might be more help if the expedition were French, or even Franco-American. This idea seems to have suggested a closer collaboration with the French and even the thought that some funding of the enterprise might be obtained from NATO.

## On to Mauritania

On June 12 they left for Dakar via Marseilles and Gibraltar, and then flew along the African coast, which made it impossible to judge anything of Saharan weather. Clouds were left behind in Spain, and Morocco was perfectly clear, but somewhere south of Casablanca the notorious desert haze began. Its white absorbing clouds were very unevenly distributed: sometimes so high and dense as to blot out the horizon, others less so, with a tendency to thin out towards the coast.

During part of the flight in clear air, they flew over an awe-inspiring seemingly endless area of desert sand. "an ocean of pink sand beyond description": "On the grand scale the sand is streaked out and stretched into row upon row of tremendously long narrow hills, all lying parallel to the trade winds."

The final approach to Dakar was over the ocean, where there was a dense sea haze. They compared the airport, not unfavorably with the (then) Austin airport, and were soon on their way for the 50-minute flight to Nouakchott, the plane being well ahead of schedule." From the air the town looked like some hard-bitten desert town, way to hell and gone out in some spot in the California or Nevada deserts, but close at hand it looks picturesque, native costumes everywhere. Dust covers everything and even in the moderate wind it is drifting across the road ". After checking in at the hotel, Bryce went out with his photometer for a sky-brightness measurement, which he found milky, not blue. His investigations were witnessed by half a dozen small boys in native dress. They were most friendly and interested in his procedures, and were soon joined by three older boys, two from the Lycée and one from the Ecole Normale Supérieure. A conversation in French ensued in which the importance of the impending eclipse was intelligently discussed.

At the hotel they were joined by Mr. Danabja, the travel agent with whom they were to be dealing. This gentleman gave the impression of being the commercial arm of the government's Ministre du Tourisme. He took them to a quiet tea-house bar, where they had a long discussion about the government's plans for handling eclipse visitors. Mr. Danabja was still at a loss to know how to handle the four or five thousand visitors expected. He divided them into two categories—those bringing all their own material, food, transport and everything—and those who would depend for many things on local entrepreneurs such as himself. He claimed that he could arrange for transportation, housing (in tents), water, electricity, etc, for fifty people at Chinguetti, and that light trucks could bring in equipment there, even though the road was not very good.

Mr. Danabja said that he had planned to meet them at the airport, but had missed them because of the plane's early arrival. He was originally going to take the DeWitts to the home of a local family, where they were invited to stay. He took them anyway, to say hello, and they turned out to be the family of Dr. Ba, the uncle of Alassane Sy. They were invited to stay for supper, and it was arranged that they should move in on the following day. Dr. Ba drove them back to the hotel, but first took them a few miles along the road to Akjoujt to see the desert sky at night, which revealed itself in all its starry splendor—'but of course it is the daytime that interests us'.

It was clear that Dr. Ba was a most distinguished Mauritanian citizen, but it was evident that they did not, at first, and perhaps not even later, appreciate just how distinguished he was. After independence in 1960, he was, for a time, Minister of Health coupled with some minor departments, during which he represented the country in negotiations with the, abortive, United Arab Republic, and came into contact with the Secretary-General of UN, U Thant and John F. Kennedy. He was later Minister of Finance, but after a policy disagreement with the President, he returned to his medical practice. He was most generously hospitable to the American expeditions, a source of the most important contacts, and always warm and genial. (Gerteiny, 1967)

Next morning, Tuesday June 13, Mr. Danabja came at 10 o'clock and drove them to the Secrétariat d' Etat pour Tourisme, where they presented their letter of introduction from the Mauritanian Ambassador in Washington to Mr. Ahmed Ould Die, the Secretary General who turned them over to his assistant, Mrs.Ould Cheikh Abdallahi, who, of course, had represented the Mauritanian government at the Madrid meeting, and her secretary, Mrs. Jacqueline Ritter. Of these two young ladies, one was English and the other French or Belgian.

Figure 20: Desert dress styles, Gentlemen of the
oasis in their bou-bous.

They completely overturned the picture painted by Mr. Danabja, scoffing at the idea that he could handle a party of 50 at Chinguetti, though they then said that the problem of housing 4000 people in Mauritania was beginning to appear soluble and that they were getting it under control. Certain regions near Akjoujt and Atar were being set aside for the 'real' scientists, and barred to the general tourists and the amateurs, who would get other areas set aside for them. All the groups must bring everything they needed with them, except food and gasoline (which *had to be* purchased locally) For customs purposes everything must be carefully documented to avoid duties, and everything left behind must be a gift, not sold.

The talk then turned to their intention to visit Chinguetti and other projected sites in the Atar-Akjoujt region. Mrs. Ould Cheikh Abdallahi had been with Bill Curtis on his ill-fated trip to Chinguetti, when the axle broke on their Land Rover. She repeated Curtis' comment that the "road" to Chinguetti was terrible, and that although possibly 4-wheel drive trucks could make it, it would be difficult to imagine how delicate apparatus would survive on a trip capable of breaking a Land Rover's rear axle. Mr. Danabja's comment to this was that the driver was no good, and didn't know when to get off the rocky spots and onto the sandy ones. The ladies recommended that they present themselves to a Mr. Lacombe, who would take care of all their local travel needs. This suggestion, made in his presence, infuriated Mr. Danabja, and precipitated the DeWitts into the midst of a local squabble. Mr. Danabja was *not* the commercial arm of the government Tourist Office, but he *was* the first indigenous tourist agent in Mauritania, and had been in operation for about four months. The previous day he received his import license and permit to handle problems for visitors. This represented a serious threat to Mr. Lacombe, an old-established Frenchman, who for years had handled tourists, importation of supplies, retailing of imported goods, etc., in Mauritania. It was clear from Mrs. Ould Daddah's recommendation of Danabja, and comments made by the Ba family, when the DeWitts discussed the matter with them, that Mr. Danabja had the support of the local population. Nevertheless it seems that government representatives, out of habit, had been automatically directing all eclipse matters to Mr. Lacombe. It appeared a bit ironic to Mr. Danabja that eclipse arrangements were effectively being made by two Europeans, one of whom, Mrs. Ritter, had tactlessly actually asked Mr. Danabja to direct the DeWitts to Mr. Lacombe. As Bryce remarked, "We have decided that since we were, for the moment any rate, free agents and not representatives of the NSF or any other arm of government, we shall deal with Danabja. We shall regard it as a test to see how he performs."

After lunch with the Bas they had the traditional three glasses of tea in the "reclining room", (Bryce could think of no other description for this nomadic-style living room), during which time two visitors arrived to see them: Mr. Mohammed Ould Khliel, Directeur de la Sureté Nationale, and Mr. Yahya Ould Menkous, Gouveneur de la Sixième Région, Rosso. (The Sixième Région includes Akjoujt : Atar and Chinguetti are in the Septième Région). They spread out the Michelin map and described their present plans and future hopes. Both men, as also Dr Ba, were dressed in light blue boubous— traditional robes—and were barefoot. Conversation was punctuated, not only by manual gestures, but also by the wriggling of toes. (Figure 20).

After tea, Dr Ba drove them to the French Embassy where they spoke to Mr. Pieton, Conseillier Culturel. Cécile described her tentative plans to lecture at the Ecole Normale Supérieure and, her hopes that the French government would pay her travel expenses. Mr. Pieton had been eager to get some cultural activity going, such as a geology conference, and listened attentively to what was said about the eclipse, in spite of its abstract nature and lack of direct applicability to terrestrial affairs.

From the French Embassy to the American one, where the only official currently present was a Third Secretary, who had arrived only ten days before. He produced a letter addressed to the DeWitts from Bill Curtis, who had left it for them on his departure from Mauritania.

In summary, this said that in 8 days he had never seen a good sky in Nouakchott, and that, of Akjoujt, Atar and Chinguetti, the last had the best sky, though this could hardly rank as a good statistical sample. His private opinion, which he asked to be treated as confidential, was that Chinguetti would be the best choice, though this presented severe logistical problems. "The road is atrocious, the air strip 7 km from the town, which has no electricity or readily available communication system with the outside world. It could support thru its charming encampment say no more than 15 persons. Most everything would have to be flown in."

Bryce was pleased to get this and certainly confirmed Curtis' assessment of Nouakchott's climate by his own descriptions of daily sand-storms and a sky full of widely scattered sunlight.

Their hosts, the Bas, both black, he a medical doctor she, a native of the Caribbean island, Guadéloupe, a teacher at the local Lycée, had a most luxurious establishment, with many servants. He had a library of rare books and kept three desert gazelles in the backyard. The DeWitts had a whole side of the upper floor to themselves. This was in sharp contrast to a great area across the street covered with the tents of the poorer population. In a note, inserted by DSE, it was thought that many of the dwellers in the tents with their characteristic 'upside down wine-glass stem' stretched canvas roofs, were in fact, nomads who had come in from the desert to escape the severe drought, which had already lasted several years.(Figure 21)

Mr. Danabja arrived after supper and Bryce got out his Questar to entertain the company with a view of Jupiter.

Figure 21: Nomads with their characteristic tents
had come into town to escape prolonged drought.

Figure 22: Desert foxes left little of a foundered camel

Next day, Mr. Danabja called with the news that he is definitely one of the government's representatives. After changing money and doing some shopping, they went with Mr. Danabja to visit a friend of his in his strictly Arab-style house, where they sat on the floor on mats in a dark room and had the usual formal tea ceremony. They then saw the director of the Ecole Normale Supérieure, Mr. Mohamed El Modar Ould Bab, and the director of studies, Madame Samuel who suggested that Cécile should give a series of seven lectures in January.

After that, they met Madame Turkiya Ould Daddah, who was very friendly. Bryce remarked that Cécile had not told him that she was so youthful and voluptuous. Her husband, who is Ministre de l'Equipment, would get in touch with them on the next Monday to discuss their plans.

They finally left for Atar at 5.00 p.m. in the usual afternoon sand-storm. However, about 30 km from Nouakchott the wind suddenly dropped, the sky cleared, and the temperature rose from 75° or 80°F to 100°F. Behind them there was a wall of dust churning in the air. It took 5 1/2 hours to get to Atar and, says Bryce, "I wouldn't recommend anyone trying to do it faster. As far as Akjoujt the road is hard-surfaced and thereafter only graded. There was rattling and banging at top speed over the washboards and sand traps of the graded stretches. Truck drivers with delicate equipment would have to go very slowly." Bryce rated the road a little better—at least straighter—than the one which crossed the Big Bend National Park from east to west close to the Rio Grande. The ground visibility on the road to Akjoujt was excellent, the sky blue except near the Sun, where it was white. They got to the government hotel at 10.30 p.m. There was no running water to get rid of the dust, but there were mosquitoes. Alas they had left their insect repellent back in Nouakchott, but they were taking anti-malarial pills. The electricity went off at midnight.

Thinking back on the trip Bryce remembered how flat the country was, with areas of scrub and short trees, and even a stretch of some twenty miles of grass where it had rained a few days before. They had seen occasional camels, and even two or three camel skeletons by the roadside (Figure 22). In the night a long-eared creature like a California jack-rabbit appeared.

Next morning there was a continental breakfast, and the hotel "gardien" was persuaded to turn on the pump so that they could take showers. Bryce set up the Questar for a seeing observation, and, when the driver came with a 404 Peugeot, tried to run the Questar drive off the car battery. This failed, apparently because of some damage to the instrument, but he did get a seeing estimate at highest magnification of between 3.7 and 3.8 seconds of arc. There was always a breeze during the day and, every so often, dust had to be wiped off, but later in the day, a test of the sky showed it remarkably free of dust. However, everybody said this had been a cool June and it had rained during the last few weeks.

While Bryce was so engaged, Cécile went to see the Governor of the seventh region, Baham Ould Mohamed Laghdaf, who received her cordially and sent a telegram ahead to his subordinate in Chinguetti asking him to facilitate their arrival. This was an old friend, Mahmoud Ould Amar Cheine. The Governor was a little concerned that the DeWitts were going in a Peugeot rather than a Land Rover, and said that the office of the Préfet at Chinguetti would put one at their disposal. Then they paid a visit to the pharmacy, where they were told that were no dangerous mosquitoes in the region, and the bank.

Bryce mentions that the method of sending messages between Atar and surrounding prefectures was via the Réseau Administratrif de Communication, a kind of radio-telegraph system, open only for a few hours each day.

Evidently any eclipse teams would have to bring in their own communication systems for contacting Atar or Nouakchott, which would require a government permit or license.

At 11.45 they went to Mahmoud's house for tea. They had not previously realized the poverty in which his family lived—a mud and brick-walled timbered roof building with a bleak courtyard, also housing a collection of close relatives. They entered a room with an earthen floor covered with woven mats. Mahmoud had two daughters and two sons, one of whom aged about 6 or 7 slept the whole time stark naked on the floor. The lad was uncircumcised though Mahmoud is a good Moslem—the previous day during the drive he had stopped the car just after sun-down and spent some 15 minutes on the sand, at some distance from the car, facing eastwards towards Mecca and praying.

"Tea" consisted of the usual three glasses, which Mahmoud prepared himself, plus the sweetened goat's-milk mixed with water from a goatskin bottle, prepared by Mahmoud's wife. The milk was served in large enamel mugs and they were obliged to drink nearly all of it in spite of their reluctance. Not that the milk was unpleasant: it was cool and refreshing though beginning to turn sour, but with all the flies crowding around they were a bit concerned with hygiene. Mahmoud gave them a small wooden bowl of the type formerly used exclusively for drinking goat's-milk. However, it then appeared that the bowl really belonged to his charming little daughter aged 4 or 5, who wept at having to part with it. They solved this embarrassing predicament by giving her one of the kaleidoscopes, with which she seemed very pleased.

Mahmoud returned them to the gîte d'étape for lunch and then came again at 3.30 for the departure to Chinguetti. Before leaving Atar they visited the barracks, which had been built by the French 15 or 20 years before. At the barracks the guard had not received the Governor's order to let them in, but let them in anyway. He and Mahmoud began by exchanging the

Figure 23: Rugged country near Atar.

Figure 24: Mahmoud, the DeWitt's
faithful driver at his home.

ritual greetings which lasted some 45 seconds, with exchange of the customary words, "God is great", "Peace be with God", and so on. Cécile said that when Mahmoud appeared before the Governor the ritual went on for several minutes.

The barracks were completely abandoned and stripped of all fittings such as plumbing and even some of the electrical conduits, but the structures were sound and could be cleaned up to accommodate 200 scientists—more if there were a good deal of doubling up.

According to Mrs. Ould Cheikh Abdallahi (Jean or the "English Lady" as she is known in Nouakchott), the French have definitely *not* pre-empted the barracks. The government would not permit this. All scientists and amateurs would be permitted to use the barracks provided their governments helped in getting the place in order. It seems this lady is highly respected in Nouakchott and Mr Danabja did not hold her in any way responsible for the behavior of her secretary in the Lacombe affair. She knew he was a travel agent, but did not know that he was handling transportation and tourist trips to the interior in addition to the usual airplane business.

The trip to Chinguetti took about 3 1/2 hours. The horizontal visibility at ground level was as good as anything likely to be seen anywhere in the US Southwest, exceeded possibly only by that in the northern Rockies or the high Sierras. The terrain resembled the very desert areas of the Southwest—there are few places in the Southwest as dry as the Sahara can get.(Figure 23). The road was somewhat worse than the one in the Big Bend mentioned earlier, and wound through a canyon and up an escarpment at a place called the Amogjar Pass. This was quite impressive, not because it was particularly high, but by comparison with the tiny road. During the drive, Bryce held the Questar in its case to cushion it against the bumps and any further damage beyond what it had evidently sustained in the pounding of the car the day before. In one place the car got stuck and they had to push,

Figure 25: Chinguetti fort, now a resthouse, gîte, looked like a Hollywood film set.

Figure 26: Wherever Bryce went a crowd of curious youngsters followed.

and, in another, the muffler fell off and was re-attached by Cécile with some nylon cord she had.

They arrived at Chinguetti just after sunset. The setting was right out of Hollywood: the oasis, the sand-coloured buildings, the palm trees, with the enormous dunes forming a back drop.The Michelin map suggested the presence of mountains, but there were none, only the end of the vast sand-sea, the *grande erg*. (Figure 25). After unloading their baggage they climbed to the roof of the gîte to view the town, while Mahmoud parked in the court-yard. The gîte had been a fort, with battlements in the style of Beau Geste. After dark they lay on mattresses in the courtyard, saw a passing satellite—which Mahmoud immediately recognized—and had couscous for supper made from one hunk of lean camel meat, with plenty of semolina (which had sand in it), but no vegetables.

While eating they heard drums and singing in the distance, which Mahmoud explained by the fact that it was a Thursday evening, and hence Friday morning, (though he called it Friday evening—he evidently reckoned the day as beginning at sun-down), and therefore the occasion for singing praises to the prophet. After a look at Jupiter through the Questar, they walked to the scene of the singing, where some 50 or 60 persons in a tight circle were singing, swaying and clapping their hands in a complicated rhythm kept by a tom-tom. Every now and then one of the women let out a strange high pitched ululation, used in former times as a call to battle.

After the singing was over, the young boys crowded round and asked all sorts of questions. The boys knew all about astronauts and events such as the death of Mao Tse Tung, with the older boys interpreting in French the questions asked by the younger ones. (Figure 26).

Next morning, before sunrise, Bryce made photometric measurements of the zenith sky, and repeated some in the afternoon. He found the visibility excellent, the sky readings very consistent and the seeing in the range of 4 or 5 arc seconds, though some of this may have been due to the wind shaking the tripod. The temperature was hot—up to 35°C (95°F), but quite bearable, although the whole town slept during the hot hours. Conditions were better than he had encountered on the roof of the Department of Astronomy back in Austin. The Percepteur (treasurer) of Chinguetti came in the middle of the morning to put the Préfet's Land Rover at their disposal. He was much pleased with being shown the sunspots with the Questar which was still set up.

The DeWitts asked the driver of the Land Rover to head out to the south as far as it was possible to go. As a result the government of the 7th region was out only five minutes worth of Land Rover time since it was after this time interval that they came to the first big dune. The Percepteur said that the only vehicles capable of crossing it were specially designed trucks with very wide tires, and if they wanted to go further they would have to ride camels. Just for the hell of it, they did this for a couple of kilometers. Bryce's camel was led by a small boy who strode barefoot over the hot sand. The camel ride proved not too bad, but the saddle was most uncomfortable when couched to permit riders to mount or dismount, the camels made an awful noise 'like a cow with bronchitis'. (Figure 27). Out in the dunes land-marks disappeared, since they looked quite different from different vantage points. At the edge of the dunes they encountered some women with two very small cute little animals known as fennecs, which had large triangular faces their ears forming two of the vertices.

Back at the gîte, for lunch of sardines and couscous, the desert ride had made Bryce extremely thirsty and he abandoned all thoughts of drinking only bottled water. For one thing, there wasn't any: For another, the electric pump was broken, so there was no running water. Tooth brushing, showering and flushing the toilet (the gîte had flush toilets !) was by buckets of water. After an afternoon sleep, they went out at about 5.30 and walked with Mahmoud to a well in the middle of a palm grove, where the water was drawn in a leather bucket dangling from the end of a long counterweighted pole (a *shadouf*). (Figure 28)This affair was worked for long hours by a man and a little boy, the latter standing on the top surface of the pole and shifting his weight back and forth as necessary, to fill the water-buckets of the young girls who came to the well. Many of these seemed frightened when they saw the DeWitts, and the smallest ones ran away. The water surface was about 20 feet below the level of the ground. It was used, not only to irrigate the palm grove, but also a small vegetable garden, a few cotton plants, and some tobacco plants, with which the inhabitants had decided to experiment the previous year. They asked Bryce how to prepare the tobacco leaves for smoking but he was not able to tell them much beyond generalities.

They were frequently followed by groups of children who demanded gifts (cadeaux), and seemed a potential source of trouble to an eclipse team. Cécile suggested that they might tell the school teacher that cadeaux would be given out on the last day of school (around June 20) the traditional school day for *'distribution des prix'*. From the well they were invited to the house of one of the aristocratic youths of the village, and went there accompanied by the little boy who had led the camel to whom they had given a kaleidoscope. This he carried in the pocket of his boubou, (located on the chest and deep enough to carry quite a number of things). The father of the aristocratic youth was away, and the son took over the rôle of host, ordering his mother and her women-folk to prepare tea. While this was going on, the camel boy grandly permitted the other children to look into his kaleidoscope. There was a definite pecking order about this, with the boys with darker skins getting the short end of the stick. Bryce did not stay for tea, but went off to make sky-brightness measurements from the roof of the gîte, where he would not be interrupted by the children.

Figure 27: When couched, camels make a noise
'like a cow with bronchitis'.

After supper of couscous, this time with mutton and perhaps a tomato skin thrown in, they set up the Questar in the court-yard for a sky show. The host of the afternoon had been asked to bring only the older boys but a number of small fry showed up, and knew exactly what they were seeing when shown the Moon under low magnification. On higher magnification, for the benefit of Mahmoud, the Percepteur, and the school-teacher, the instrument was turned on Jupiter, showing for Mahmoud the changes in the positions of the satellites from those he had seen the night before. The school-teacher was especially interested in Saturn's rings and Bryce gave them quite a sophisticated account of orbital motion and gravity. The implication not explicitly stated by Bryce, but borne out by Evans' later experience, was that these people were highly intelligent, seemed well-nourished, in good health, and apart from the modest pressure of the children, extremely well behaved, with their elders most courteous and generous, but, except for the emancipated wives of high-ranking individuals, essentially a male society, with women subject to many of the restrictions imposed by Islam.

Bryce summed up their experience in Chinguetti, in his diary entry:—That a team there would have to accept the fact of being 15 kilometers north of the eclipse track center-line: that the town could not support many people. Not more than 20 could be accommodated at the gîte, and nobody should come who was not prepared to live rather primitively, either on C-rations or local couscous. Safe drinking water could be trucked in—maximum truck tonnage 5 tons—from Atar or local water could be purified. The best kind of structure for housing the telescope would be made of local stone and cement or mud. Roofing like that of the gîte would be of palm logs, and a generator would have to be brought in to supply electricity, though Bryce remarked that the Republic of Libya was supposed to be bringing in an electric generator and making a new road to be ready by the next year. The opinion seemed unanimous that sandstorms produced less dust in

Figure 28: A Shadouf at Chinguetti,
the traditional equivalent of a water pump.

the air than at Atar or Akjoujt, supposedly because the sand was coarser than the soil at those two places.

Next day, Saturday June 17, Bryce was out early to make sky—brightness readings, and they left Chinguetti immediately after breakfast taking with them the Percepteur, who was going to Atar to see a doctor. Just before leaving they were approached by the gardien to ask them to change into CFA two American Express travelers' checks, for $10 and $ 20 given to him two weeks earlier by Gene Prantner from NCAR in payment for a balance of Fcs 7500 on the bill for himself and Jean Ould Cheikh Abdallahi for their stay on the night of May 24. The gardien showed them his bill and they agreed to make the exchange, but made the mistake of discussing the matter with the Percepteur, to ask what the gardien could have done with them, particularly when they pointed out that Prantner had omitted to countersign them. The DeWitts knew the checks were no good to them, but were confident that Prantner would later reimburse them.

On learning of this complication, the gardien appeared to have felt that he had been taken, for he suddenly stopped talking. He had thought the checks were simply money. On top of this the DeWitts found that they only had enough CFA to pay their own bill, and only had French francs to exchange for Prantner's checks. The gardien evidently decided that it was safer to hang on to the checks even though he had no way of converting them to Mauritanian currency. The DeWitts proposed on their return to Nouakchott to go to Manger at the American embassy to do what he could for the gardien.

On the drive to Atar a steady wind from the northeast got up, a regular trade wind, which brought some dust and diminished visibility. At Atar they went directly to the Governor's office to thank him for the hospitality extended to them at Chinguetti, then to the market to get something for lunch. Finally to Mahmoud 's house for a five-hour sequence of sweetened goat's-milk, tea, grilled meat and slabs of fat, more tea, couscous, praying, singing and dancing by the children and then more tea. By now they were thoroughly at home in the Mauritanian style of life, sitting on the floor, small windows at ground level for ventilation, eating with their hands, washing hands after meals rather than before. By mid-afternoon the Sun was still only a two-finger Sun, the rough-and-ready way of assessing atmospheric scattering by seeing how many finger widths it took to hide the full Sun. The Questar had been set up at noon giving the family a look at the sun-spots, followed by a seeing estimate of 5 or 6 arc seconds using the long eyepiece. Is the steadiness due to the regular breeze that blows in this region? Bryce had earlier remarked on Mahmoud's poverty, but in fact, by local standards, he was quite well off, owned his own car, which he used for transporting people, and a date-plantation to which he took his family for vacations,.

They left Mahmoud's house at about 4.00 p.m. after warm embraces between Cécile and the family, and tears from the littlest daughter when she saw her father going off again. At the edge of town they were stopped by a policeman who told them to wait until one of his kinsmen showed up in a bus and ordered them to take him to Akjoujt. Mahmoud had brought with him a companion who knew the area to help him find the way to El Loueïbda, a place near Akjoujt that they wanted to visit, so the policeman's relative, Mahmoud's friend and Bryce, with the Questar on his lap, rode in the back together. Mahmoud's friend, an older man, had a high-pitched monotone voice, and half the time chanted prayers to himself. His regular speaking voice had the same quality.

The road between Atar and Akjoujt, which they had covered in darkness before, was seen to be rather pretty in spots—hills and, rarely, occasional palm groves. At one point they stopped at a well while one of the men crawled down inside it to fetch water, bracing himself on the walls like an alpinist in a chimney. The wind got up and the visibility decreased but improved when they got near Akjoujt. Blowing sand drifted across the road, but things were not as bad as during a regular sandstorm that Bryce had encountered in the previous September crossing the California desert out of Yuma. He would very much have liked to know whether Chinguetti was simultaneously experiencing decreased visibility. They had been told that Chinguetti had had a sandstorm between the time of Bill Curtis' visit and their own, and if it were again having one, that would make an average of 15 days between storms, though this sort of averaging might be nonsense.

The weather got hotter near Akjoujt; Mahmoud said that this was because Akjoujt was lower than Atar, in its turn lower than Chinguetti. The men in the back seat, and Mahmoud himself, would from time to time spit out of the window, as if to clear their nasal passages of dust, though it was not clear whether this was related to the Moslem practice of throat clearing and spitting in the early morning, of which Bryce and Cécile had had an excruciating experience in India in 1951.

Later, the sky became more diffuse, which should have been apparent in a photograph of the setting Sun taken just north of Akjoujt. While the car was stopped the Mauritanians took the opportunity to say their sunset prayers. Arriving at Akjoujt at dusk the DeWitts had no idea where they would sleep. Prepared to sleep under the stars, the idea didn't seem so attractive, even with the diminished night wind expected, in the prevailing dusty conditions. Mahmoud offered to put them up at the house of one of his relatives, which would have meant sleeping in a courtyard or in a room with 5 or 6 other people, but he eventually decided to take them to the SOMIMA camp just in case there might be something available. It turned out that there was just one two-bed cabin available. Mr. Frances Guichot, the "gérant" of the restaurant received them graciously. He and his wife were slightly astonished that the DeWitts had been out in the back country with a native chauffeur, exhibiting a rather comic confirmed colonial attitude. In spite of being in Mauritania for 18 years, they had never been to Chinguetti, nor to any other of the interesting places. The camp was very insulated from Mauritanian life. The young French, American and Canadian employees, who lived in the camp, rarely visited outside. After a shower in the bathhouse, changing to civilized clothing (into summer street clothes from the previous rig of jeans etc.), they had two beers and a French style supper, before going to bed in their air-conditioned trailer.

At this point in his diary Bryce introduces a series of paragraphs summing up his recollections of the Mauritanian trip. He lists the animals seen, apart from goats and camels, starting with an eight-inch yellow lizard which swiftly buried itself in the sand to escape capture, as well as some large black lizards. A marmot-like creature in the Amogjar pass canyon: two fennecs (desert foxes) : large scavenging black birds: a kind of hawk or falcon :and ground-running birds like quail, which emitted a pleasant chirp-warble. Mahmoud temporarily captured two by running after them: three caged gazelles.

Star names : Although most modern bright star names are of Arabic origin the modern Arab does not know them. There is a library with many ancient manuscripts at Chinguetti, and they had heard that Chinguetti was an old "university" town. Mahmoud knew of this, but not of the library, which the Governor of Atar told them about later.

Couscous is flavoured with onion and tomato peel—the Arabs dislike the flavor of whole tomatoes—goatbutter, a yellowish liquid, is added for flavor and consistency.

The goatskin *guerba* (water-container) keeps the contents astonishingly cool by evaporation.

Dress: Moslem women above a certain age hide their hair from all men except their husbands with a fold of their robes, leaving clear only a tuft in front. The back hair is kept short. Mahmoud's wife became quite casual with us and one could sometimes glimpse a sort of pig-tail behind each ear.

Her face was intelligent, motherly, affectionate and wise, but the alleged erotic significance of the hidden face was lost on Bryce.

Mahmoud smoked a tiny pipe with a bowl, which pointed in the same direction as the stem, for about a minute at a time—which exhausted the tobacco—during rest-stops or other moments of relaxation.

The Mauritanian toilet is a high platform of logs with spaces between, surrounded by a low wall to hide the occupant squatting with one foot on each log. The extreme dryness makes this odor-free, in contrast to the European-style toilets in the gîtes d'étapes. The DeWitts could not determine the age of toilet-training and noticed, at one time, an 8 or 9 month-old child crawling around in the sand among the guests leaving a trail of moisture which occasioned no comment.

The Questar: At Chinguetti Bryce had tried to get into its base to see why the hour-angle drive was failing, but had been unable to do so. This was presumably some kind of friction drive, which had suffered from rough handling, but it had started to work better again.

Next morning, Sunday June 18, Mahmoud called for them at 7.45 at the SOMIMA camp to take them into a region north of Akjoujt that La Count had said in Washington, might be one of the eclipse sites. Bryce had not kept a copy of the map he had been shown, but from memory he judged that the place in question would be a well, known as El Loueïbda. The only road worthy of the name in that region was the one that led to the well, and even that was unknown to many of the local people. The local gendarme gave them some starting directions, which proved to be false, as the road soon disappeared, and they wandered around until they met a man on a camel, who pointed eastward, but they only found the road after an encounter with a third such gentleman. The desert may seem to be empty of life, but in reality is full of gentlemen on camels. In the mean-time Mahmoud expertly guided his

404 Peugeot across one flat, rise, shallow stream-bed, or sand-trap after another—a superb advertisement for the Peugeot people. The road, when they found it, was not much more than a track running between two lines of small hills, but the desert was so smooth and flat along its course that trucks could easily carry delicate equipment over it without difficulty. The well was about 32 kilometers from Akjoujt at the end of the road. They stopped there, while Bryce set up the Questar for a seeing observation, and Mahmoud's friend, who had proved quite useless as a guide, ran off to find a few small pieces of wood to make a fire for tea. Although the wind had died down in the night, it had got up again, and it seemed impossible to make a fire for the tea, but this was achieved and on it the tea-kettle boiled merrily. Some scattered altocumulus clouds were covering the Sun just then, and it took two glasses-of-tea-worth time for them to move on.

In spite of a slight haze and visibility down to about ten miles, the seeing was remarkably good, perhaps 3 arc seconds, about the best that could be done with the Questar. The wind had got up a bit, blowing sand, and Bryce got everybody to hold a blanket as a wind-break.

Near the well was an apparently abandoned cistern, quite empty, divided into three 11 x 33 foot rectangles, with sides about 3 feet high, which might be useful as a telescope shelter. They then headed back to town and on by the hard-surface road to Nouakchott. He noted that El Loueïbda was about 12 kilometers south of the central eclipse line, as compared with Atar, which was 40 km north, and Chinguetti, which was either 13 or 18 km north, depending on whether the town or the airstrip was chosen as the site for observations. Just outside Akjoujt, Mahmoud stopped for gasoline at the only station in that town. The pump operator was absent and nobody else had the key. Mahmoud said that this was one of the hazards of his occupation and that, sometimes, even when the pump operator was there, there was no gasoline to be had. He had a jerrican of gas stashed away near the station for such emergencies, though he hated to use it, since there was no guarantee that a supply would be any more likely the next time he passed that way. Everybody knew about the can, but such is the honesty of the average Mauritanian that nobody had touched it.

They were on their way again by about 11.00 a.m., by which time the wind had switched round to the north, and a real sandstorm began.It got very hot, up to 42°C (108°F) by 1.00 p.m. Sand swirled across the road and red dust was thick in the air, at first only in patches with good visibility and some superb mirages, but then the visibility decreased, at first to 2 miles, then one mile, and finally to less than 100 yards. Overhead, the sky could occasionally be seen, with a widespread layer of altocumulus. Bryce asked whether these clouds naturally occurred with sandstorms, to which this later commentator would answer in the affirmative, this being the Intertropical Convergence at work. There was no on-shore breeze and Nouakchott was hot. Mahmoud dropped them at Dr. Ba's house, in time for a quick shower and lunch, buffet style. Another guest was the Senegalese Ambassador to Mauritania, and Bryce and he and Dr. Ba, who mixed a taste for philosophy with science fiction—he was a fan of Jules Verne—discussed human destiny, while the women talked of more practical matters. The rest of the day was devoted to resting and making a few phone calls.

Bryce was so exhausted that he ran the entries for the next two days, 19 and 20 of June, together. On the first day they visited the American Embassy, and saw Mr. Manger, one of the secretaries, and Ambassador Murphy, who was still worried that Mauritanian officials did not have eclipse arrangements well under control. Considering that he was presumably the official conduit of the US Government to the Government of Mauritania, it was remarkable that he should advise the DeWitts to keep all options open, both private through Mr. Danabja, and officially through the NSF.

They then went with Mr. Manger to see Jean Ould Chekh

Abdallahi, who was very interested in their experiences at Chinguetti and how well Danabja had served them. Bryce asked her whether they could be sure of getting Chinguetti if they wanted it, to which she replied that considerations of international cooperation would make it impossible to operate on a first-come-first-served basis, i.e. everyone would be a VIP. He was most concerned to realise that she made no distinction between "bona fide" scientists and, for example, Menzel's amateurs, not that this was a slur on them, but distinguished them from those who would publish their results in professional journals.

Cécile came in near the end of the discussion and Mrs. Abdallahi was most interested to learn that she would be giving lectures in Nouakchott in January. This cheered her up no end. It was clear that what Mrs. Abdallahi cared about most was keeping people happy.

Cécile had been to the bank to get some francs to pay Mahmoud. He had gone to the Ba's house at 8.30 as previously arranged, but by some mischance was informed at the gate that the DeWitts had already left. So he waited there while Cécile waited for him inside, and only Bryce's emerging from the house to go the Embassy resolved the mix-up.

Danabja's office called to say that their flight to Dakar had been cancelled owing to the world-wide airline strike. They would have to drive to Dakar to catch their plane the next day, and this meant leaving at 3.00 p.m. to get to the ferry across the Senegal river before it closed. So everything got very rushed, and though they had hoped they could drive to Dakar with Mahmoud, Danabja put his agency's chauffeur-driven car at their disposal, at a considerably lower price than Mahmoud could afford to offer. Mahmoud was a very good sport about it, for Danabja had put business in his way—and be cause he is a high-class Moslem and that is the nomad's way. Because their whole trip had taken a bit longer than Mahmoud had originally planned, (although no mention had initially been made of time), Cécile added a small supplement to the agreed price. He did not ask for it.

Cécile then went to report to the personnel at the French Embassy concerned with cultural cooperation with Mauritania, that arrangements for her lecture series had been completed. Bryce thought that this paved the way for possible French support for the eclipse effort if any sort of collaboration with the French could be developed.

At the bank, Cécile had been given priority treatment by Mr. Prié, husband of Danabja's secretary, and moved to the head of a long line. The DeWitts felt that Danabja's performance on their behalf was superb. The total cost of their transportation to Chinguetti and back was Fcs 69,000 CFA ($ 345), about half to three-fifths of what it would have been with one of Lacombe's Land Rovers, much more uncomfortable than the Peugeot, which, with Mahmoud at the helm outflanked the sort of sand-traps that had proved too much for the Land Rover driver supplied by Lacombe to Curtis and Prantner. They learned later that, when that driver got stuck in the sand, he had tried to bull his way out with the 4-wheel drive until the whole vehicle was in so deep that the axle broke. Another item—Lacombe could never have arranged for them to stay with the Ba's—free. Brahim Danabja (to give his full name) was a bachelor, retired policeman and hunter with all sorts of contacts in Mauritania. They could recommend him most highly.

From the French Embassy, Cécile went to the Mauritanian Ministre de l'Equipment, headed by Mr. Abdallahi Ould Daddah, the President's brother. She was received very warmly, and he told her that he had recently placed a new boss over Ould Die : Mr. Chaïkh Abdallahi Sidi, Ministre du Développement Industriel, responsable de l'économie au Comité Permanent du Bureau Politique National. Mr. Cheïkh Abdallahi Sidi is the man who, from now on, would make all the real decisions regarding Mauritania's contribution to the international eclipse effort. Mr. Ould Die had become his chief executive aide. Quite early in the game it had become clear to the DeWitts that Ould Die was a rather poor organizer, and was only too ready to pass the buck to Jean Ould Cheikh Abdallahi, who was very nice, but somewhat inexperienced. The reason she was so highly regarded by Americans was that she spoke English, and was available. Mr. Ould Daddah told Cécile that he was planning to replace the existing broken electric generator, and to improve the road to Chinguetti by eclipse time, no matter what the Libyans did. They hoped he meant it.

While Cécile was making these calls Bryce returned to the Ba's house to meet Mr. Galluedec, a former French army pilot who had been in Mauritania for 28 years, and was married to a local woman. He ran an air-taxi service, and was the one who flew Prantner out of Chinguetti when the Land Rover broke down—after Curtis had gone back to Atar in the other one to get help. Galluedec had over 8000 flying hours in Mauritania and was widely regarded as knowing every rock in the desert, and a fount of reliable weather information. Galluedec, who was very soft-spoken, by no means scorned the west African weather service ASECNA (Agence pour la Securité de la Navigation Aerienne), and regretted that there were so few weather posts and pilot reports that ASECNA was often in error. He made several important points :

(1) Although conditions for the occurrence of sandstorms often developed simultaneously over wide areas, there was wide local variation in the details.

(2) For some reason, visibility in a sandstorm was always better over rocks and mountains, possibly because they contain more water, less sand, and less of the clay that produced such fine dust outside the sandy areas.

(3) Although the dust can reach to 5000 meters, the bulk of it tends to remain near the ground. He illustrated this by recalling occasions when he was able to follow a road by looking straight down with cars below crawling along with their lights on.

(4) The Nigérien Sahara resembled that in Mauritania, though the sandstorms tended to be more severe. This was because the Intertropical Convergence had a stronger influence there than in Mauritania. At the beginning of the rainy season the storms there tended to be localized but could develop into frontal systems later in the season.

(5) In Mauritania *the area that almost always had the best visibility was that of Chinguetti,* (he said this with emphasis), *Akjoujt was the worst.* Bryce had found Nouakchott worst, but that may have been only temporary. Galluedec thought Chinguetti was best because of the lack of clay-dust in the sand to the south and east, because of the rocky nature of the ground to the north and west, and, in particular because of its location on a plateau. Although its elevation was only 600 meters, the Amogjar escarpment at the edge of the plateau (900 metres) was the significant feature.

(6) A stone building left at the airstrip in Chinguetti might be useful. At present there was nowhere to store things out of the dust.

Cécile returned in time to meet Galluedec, who left soon after. Then there was a mad rush to eat lunch and pack before the car was to come at 3.00 p.m. to drive them to Dakar. By 4.00 p.m. the car had not arrived, and a frantic call to Danabja revealed that the driver was nowhere to be found. At this, Dr. Ba placed a call to his friend, Mr. Athie el Hadj, Préfet of Rosso, the town where the ferry is located, to have him keep it in operation until they arrived. Shortly afterwards the driver appeared and they said a hurried goodbye to the Ba's and were off.

With the impending departure of the DeWitts from Mauritania, the later chronicler takes to himself the privilege of describing the tea ceremony, already referred to several times, and likely to be mentioned often in the later record. The tea-pot is often a small, somewhat ornate, pewter vessel. Pewter is an alloy of tin and lead with a relatively low melting-point, but not hazarded by the tiny fires of sticks used to bring the water to the boil.

Figure 29: Decorated teapot, centerpiece of the
traditional desert tea ceremony.

At one time it was thought possible that everybody who habitually took part in this ceremony might be at risk of lead-poisoning, but, so my encyclopedia tells me, the lead content is kept small enough to avoid this. The tea-pot which I brought from Mauritania has an elegant curved spout with a non-drip arrangement at the tip, a spire-like ornament on the lid and some other surface decorations, and holds about enough liquid to fill an 8-ounce drinking glass. (Figure 29) The host boils the water, and then pushes in tea leaves and mint, before opening the box where the sugar is kept, brushes off the flies and adds some of that. When ready, the tea is poured into small thick-walled straight-sided glasses, hardly larger than a barman's shot-glass. Everybody gets one, and duly drinks. The glasses are then returned to the host, who wipes them out with a cloth, again pours the tea, and hands round the glasses. There is no guarantee that each guest will get back his original glass. Again the participants drink, hand back the glasses, to be wiped again, and refilled. After the third glass the ceremony is over. The dictates of sparse supplies of wood and water obviously govern this procedure, but in addition it also clearly serves as a guarantee of peace, of absence of poison, somewhat as one imagines the Indian pipe-of-peace rite ensured the same kind of amity. On one occasion, oversensitive to the bacteriological possibilities of the rite—I had had a good look at the wiping cloth—I made as if to retire before its completion, at which point it was quietly indicated to me that this would be a social gaffe of the worst kind.

## On To Niger

The drive to Dakar was long and tiring and the DeWitts arrived on the banks of the Senegal river at 6.30 p.m., half an hour after the ferry normally stopped running. Cécile ran to the customs office to call the Préfet to thank him for his kindness, to which he responded cordially. The last ferry of the day carried the DeWitts, a Volkswagen microbus and an enormous gravel truck, which together filled the rather dilapidated boat. On the other side of the river they quickly passed through customs, with Bryce showing his Senegalese visa obtained free at the Nouakchott embassy as the result of his meeting with the Ambassador at the Ba's. Soon afterwards they were stopped by a policeman, who asked them to take a passenger as far as Thiès, about 70 km north of Dakar. The passenger turned out to be a friendly fellow, a merchant, who guided them to a nice softdrink bar in St. Louis, where they stopped for gasoline. The passenger had two wives and five children.

On the ride from Nouakchott to the Senegal river, one of Africa's grandest, the dust in the air gradually turned to haze and the air became more humid. South of the river the vegetation increased markedly, there were thatched huts, and in the dark, occasional huge trees with spreading branches over the road. Senegal was evidently much richer than Mauritania with telephone lines, good roads, big cities, numerous industrial plants, and one of the finest ports in Africa. The passenger told them that the ferry at Rosso was out of order one day out of every two or three, because of constant breakdowns, so the DeWitts were lucky, but noted that this fact should be brought to the notice of logistics planners. The passenger was let off at Thiès at about 10.00 p.m. and an hour later they were at their hotel, the N'Gor, in Dakar belonging to a chain, Les Relais Aerien Français, to which their hotel in Niamey also belonged. The N'Gor was luxurious but modest in price, with the equivalent of about $21 for a double room, including breakfast. Danabja had chosen it because of its proximity to the airport.

They had to be up at 5.00 a.m. to catch their plane. Visibility on the flight to Niamey, capital of the former French colony, Niger, not to be confused with the contiguous former British colony, Nigeria, was generally good. They stopped en route at Bamako, Bobo-Dioulasso and Ouagadougo. This part of Africa was still very wild and empty, with vast stretches of semi-arid country interspersed with lightly forested uplands, not cut by a single road for miles. There were many shiny aluminum roofs at the towns where they stopped, as well as many African-style dwellings, compounds of thatch or mud-huts. Cécile stepped out of the plane at each stop, but Bryce felt too listless to do so except at Ouagadougo, the lunch stop, where he had little appetite.

They arrived at Niamey about 3.45 p.m. West European time. West of Niger kept one hour earlier. They were met by Paul Inskeep of the American Embassy, whose wife and small son were arriving on the same plane. With him were Douglas Burritt and Gene Prantner. The latter had in his hand a telegram received from Ambassador Murphy in Nouakchott in regard to the travelers' checks at Chinguetti. He was as mad as a hornet. According to him the checks had been stolen from him, and he had never owed Fcs 7500 CFA at the gîte. This might have explained why the gardien had stopped talking when they pointed out to him that the checks had not been counter-signed. The gardien had been very convincing, producing a record of the bill. Because of the rigorous honesty of the nomadic Islamic tradition the DeWitts had been quite unsuspicious. The explanation for this apparent deviation from tradition was the fact that the gardien was not an Arab. Bryce intended to write a letter of apology to Ambassador Murphy, regretting that, from a desire to keep the affair "in the family", they had not checked the gardien's story with Mrs Ould Chiekh Abdallahi, who had been at Chinguetti at the time.

As usual they got through customs easily even with the Questar and their supply of dehydrated food. Prantner wanted to know what Bryce was planning to do in Niger and said that he would be wasting his time going to the Aïr mountains. As he put it, he was going there to prove that it, i.e. sending an eclipse team, could not be done. Bryce explained that originally he had no intention of coming to Niger, but that colleagues in

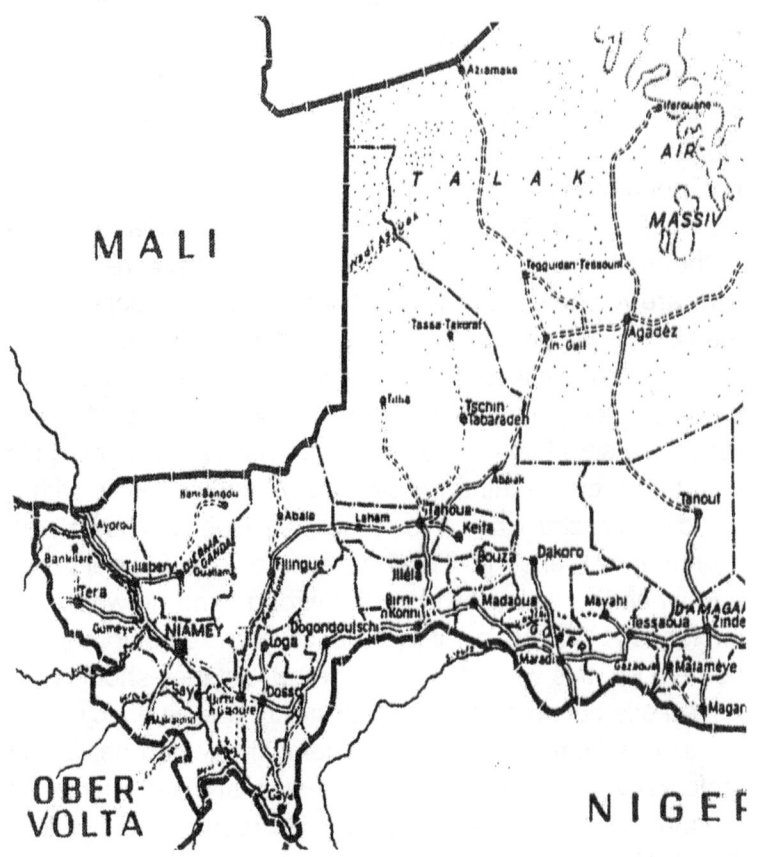

Figure 30: Sketch map of Niger.
(Adapted from Beuchtel, 1968).

Texas were anxious to have him investigate conditions in the Aïr mountains as compared with those elsewhere. (Figure 30).

After they had checked in at the hotel, which overlooked a great sweep of the Niger river, (Figure 31),Prantner invited Bryce to join the rest of them at the bar, which he declined on account of fatigue. Cécile had gone out to Air Afrique to reconfirm her travel arrangements, and then sent for some soup for Bryce, before going down to the bar, where she met Donald Ecklund and Claude Morel from NCAR, whom Bryce had not previously met, together with Prantner and Douglas Burritt. Prantner brought up a subject which seemed to be bothering him a lot. He said "I'm in charge of site inspection. Why are you here?" Cécile explained to him, with more details than Bryce had given to him in the car, the sequence of events which led to their being there. This didn't make him any happier saying, "I don't want other reports going back to La Count". Bryce remarks that Cécile, when she wanted could be quite disarming, but Prantner, a big heavy-chested fellow with a blond moustache, was not to be mollified. At the bar he ordered a drink in a French that even Cécile could not understand, and when the waiter brought the wrong thing he was quite unpleasant. Of course, many Americans and other foreigners behave like this when they are abroad, but they should not.

An example of this lack of sensitivity was given to Cécile by an American named Lichtenstein who joined her at the table. He was in uranium mining studies, and had spent a lot of time at the American Embassy. He reported to Cécile a story that had been going around among the Americans during last year's site inspection. It seems that one of the government ministers in Niamey had received a U.S. group, who were explaining their hopes and plans for the eclipse. The Minister was very cordial and at one point remarked that June was an unfortunate choice of season, adding "Couldn't you Americans change the date of the eclipse?"

The Americans gleefully recounted this as evidence of the ignorance of Nigérien officials, not realising that many of

Figure 31: At Niamey, the hotel overlooked a
great sweep of the Niger river.
(Photo. Dr. John James).

them had been educated in France, and spoke a French of greater refinement than the English spoken by many Americans. The truth is that the Minister was either making a playful joke or (equally likely) taking a subtle dig at the Americans. "Can't you Americans, with all your money and achievements in space, even affect an eclipse?".

After supper Cécile, realizing that she had only a very short time in Niger, telephoned Professor Boubakar Ba, who invited her to go to his home there and then, which she did, in spite of her fatigue. He took Cécile all over the town trying to locate some others of the academic contacts already mentioned earlier in Bryce's diary, in connection with the proposal to have Cécile deliver some lectures. He said that his university would be very grateful for any books that the Americans might send or leave behind. So far as the eclipse itself was concerned Professor Ba said that the scientists would have to bring with them everything that they wanted and hoped that most of it would be left behind, since Niger could supply nothing. This was in sharp contrast to the attitude in Mauritania where the concern was whether the locals would be able to entertain the vistors hospitably. Bryce felt that they had already been able to see a contrast between the ordinary behavior of people in the streets and the aristocratic behavior of the Mauritanian Arabs. Bryce acknowledged that they had yet to meet the desert nomads in the north of the country. Professor Ba brought Cécile back to the hotel around midnight and invited them to have supper with himself and his wife at their home the following night.

Bryce had a very bad night with an intestinal upset, reminding him of a long illness he had had in India in 1951. He felt a little better in the morning but could eat nothing. Cécile began to feel it too, so he made an appointment to see the Peace Corps doctor at 10.00 a.m. They joined Prantner's group at the swimming-pool for a few minutes and Bryce met Donald Eklund and Claude Morel for the first time. Prantner was still unfriendly and asked how Bryce thought he could ride up to Agadès, spend two days in the Aïr, and ride back to Niamey in time to catch his plane back to Paris. He evidently thought it would not be possible. The rainy season had come a month early and made the direct route to Agadès impassable, and Prantner had ordered Burritt to go by the long route, which would take three days instead of two. Bryce replied that, if necessary, he would postpone his trip to Paris, but that, at that moment, there was considerable uncertainty whether he would be able to make the drive at all, depending on the advice of the Peace Corps doctor. Burritt, who was friendly and quiet, as they all were in Prantner's presence, said that he was not going alone, and counted on Bryce's ability to speak French, since his driver spoke virtually no English. They then left, leaving arrangements in the air, but shortly before 10.00 a.m. they came upon Burritt who was arranging some things in his car, and asked whether he could give them a ride to the doctor, but he mumbled something like "I don't know", and then turned to Prantner who was standing nearby. They eventually got a taxi. The fare was only Fcs 100 FCA (about 40 c US) from any point in the city to any other.

The Peace Corps doctor, (Milton L. Kogan M.D.), was normally stationed in Ouagadougou, but made periodic visits over a wide area. He was a big balding young man with a black moustache and obviously lots of physical stamina. He examined Bryce and found nothing serious, but told him to stop taking his intestinal pills, so that if he did have anything that needed dosing it could be diagnosed. He said it would be foolish to set off for Agadès and the desert heat in the morning. He recommended a few days rest, and then to take the bi-weekly flight to Agadès. His manner was at all times that of the old-fashioned country doctor, with plenty of time for talk. He charged nothing for himself, but assessed a small fee for the 'caisse noire'. As he put it, "The Peace Corps always needs money".

From the unpretentious building of the Peace Corps, they went to the Cultural Ministry (la MAC) to see a Mr Chauvet about the possibility of Cécile's lecturing in Niamey in January, and paving the way for possible scientific financial support in case a collaboration with the French developed. Mr Chauvet was assistant to Mr Courty, one of the academic contacts given them earlier, but he was out of the country just then. Mr Chauvet remarked that Mr Idé Yacouba, Préfet of Agadès was in town just then, and made an appointment for them to see him in the morning. They then went to see a Mr Caldéron at the local office of SOMAÏR to get more information about the Aïr region. Mr Caldéron, who was, at the time representing both SOMAÏR and the CEA in the absence of other officers, was extremely friendly, and answered many questions and offered leads that they, unfortunately, would not have time to follow up on that trip.

Bryce remarked that he had found that NCAR, and, through them, NSF were very well informed about Niger. There was little that the DeWitts had been able to learn which they did not already know, except for some information from the French and la MAC, but that they were not as well informed about Mauritania. Bryce still thought it a good idea, in view of the uncertainties of the future, to keep their eyes open in case they might stumble on something of especial importance to their own team.

At the hotel, Bryce went to Prantner to say that he would be unable to accompany Burritt on his drive to Agadès, but, feeling guilty at leaving Burritt in the lurch, offered to try to get Bouchez, one of the DeWitts' academic contacts, as a replacement. Professor Ba had indicated that since university examinations were over, Bouchez would jump at the chance for a ride to the Aïr.

Bryce told Prantner of their appointment with the Préfet of Agadès, and asked him if he would like to come along. As far as Bouchez was concerned, he said "Go ahead", but as for the Préfet, said "Why should I see him here when I can see him in Agadès?", which wasn't true because the Préfet was going to be in Niamey for ten more days.

Bryce was again beginning to feel poorly and asked Cécile to seek out Bouchez, but she was reluctant to send him on an expedition, the organization, purpose, and basic chain of command of which she was uncertain, and about the congeniality of which she had strong reservations. She therefore sought out Morel as someone with whom she could converse in her own native language, with all its nuances, in order to get the picture completely straight. Morel was very friendly, but extremely careful not to express any criticism of Prantner. The message conveyed was that Prantner regarded himself as completely in charge of all the inspections for the NSF, that he was annoyed at the British report on El Meki, which they have chosen as their site in the Aïr, and that the appearance of Bryce and Burritt on the scene was a complicating factor for him. He resented the fact that La Count presented him with a fait accompli in arranging for them to be there.

Near the end of Cécile's discussion with Morel, Prantner appeared and said that he would not need Bouchez. He had arranged for Ecklund to ride up with Burritt, while he and Morel flew to Agadès. Although neither Ecklund nor Burritt spoke French, the former would be more comfortable with a companion, and would join the others camping on the road to Arlit, and then would return to Agadès to meet Bryce who would fly up on Monday.

The DeWitts had lunch in their room with Bryce still unable to eat much. After resting, Bryce went along to the hospital to get a test recommended by Dr. Kogan, and Cécile to the Air Afrique office to get new reservations for him. Because of the scheduling of flights to Agadès he would have to leave Niamey on July 1 rather than on June 29. At the hospital laboratory, although the personnel were still on duty, he was told that it was too late to have the test done, and that he would have to come back in the morning. Cécile had better luck at the Air Afrique office, and not only got new reservations, but arranged to have the unused portion of their Nouakchott-Dakar flights applied to the cost of their tickets. Bryce wrote a warning note for reference by future travelers, that domestic flights on Niger were assessed a 12% tax, which was not applied to the same tickets purchased outside the country. Prantner had not had similar good experience, and in a memorandum, had called the Air Afrique service terrible. Then to the Office de l'Energie Solaire to locate Moumorini to give him the letter and gift from Miss Adda. It turned out that he had left Niger two days before on vacation.

To supper at the Ba's, where they were met by Mrs. Ba, who was a native of the Caribbean island of Martinique. Ba himself arrived late having been tied up by a faculty meeting at the university. He is a differential geometer, and it turned out that Bryce had met him ten years before in Santa Barbara at a gathering of mathematicians and physicists organized by A. Taub and C. Misner. He said he would be happy if one of the university's physics students went along with our team but cautioned that the Nigérien students were pretty green. It was a very pleasant evening, which Bryce could not fully appreciate because he was feeling ill again. They left rather early, being driven to their hotel by Dr. Ba, the day ending with a call from Dr. Kogan, asking how Bryce felt, saying not to worry about not eating, but to drink soft drinks, such as Coke.

Thursday June 22 : It rained heavily in the night with lightning and thunder, and Bryce felt much better, but Cécile felt worse, keeping going on intestinal pills on the excuse that she would be returning to Paris the next day.

Burritt and Eklund left early, in spite of the rain taking the direct route to Agadès, and were gone by the time Bryce got up. Paul Inskeep did not go with them, being too pre-occupied with embassy business. There would thus have been a place available with Bill Curtis. The NCAR group with less expenditure could have driven to Agadès, and explored the Aïr, as well as camping in the desert to the west, and making the route survey, thus obviating the need for Burritt to be sent for this purpose.

After breakfast they went to the Air Afrique office to pick up tickets, having previously been to see Prantner to tell him that, he, Bryce, was better and would definitely be in Agadès on Monday. Prantner was in a better humor and agreed that Burritt would meet Bryce there, and would take him to the Aïr on Tuesday and Wednesday. He asked Bryce to see what he could do about getting another Land Rover in Agadès, since he did not like the idea of a single vehicle going into the Aïr. He would prefer not to have to go there himself at all, but would pay for the extra vehicle. Bryce said he would see what he could do, and asked Prantner if he could see his copy of the British report on El Meki, saying that he himself would be sending a report on his own observations in the Aïr as La Count had requested. Prantner replied that his copy was in the luggage in Burritt's Land Rover, already en route to Agadès.

Everything was in order at the Air Afrique office, and the DeWitts then went to the Ministère de L'Intérieure to see Mr. Laroya, Directeur du Cabinet, détaché à la Coöpération, and to see the Préfet of Agadès who would be arriving in his office. While waiting for the Préfet, Laroya said that an M.I.T. group would be coming to Agadès for an independent site study. The DeWitts had heard this rumor and had asked Prantner about it, but he too, had only heard the rumor. The Italians were reported to be going to Timia.

The Préfet proved to be a very handsome man in a white embroidered boubou, speaking impeccable French. He was most cordial, and said that they must not think of owing the prefecture anything in return for the disruption of local life occasioned by the eclipse. His attitude was more in the style that they had become accustomed to in Mauritania. Because of the early rains, the Préfet thought that clouds would develop in the day-time in the mountains. The Baguezane mountains could only be reached by camel, but there was water and electricity at El Meki. He recommended a better hotel for them in Agadès, and when Bryce raised the question of obtaining another Land Rover, dictated the text of a telegram to be sent to Mr. Guidoni, Director of Public Works in Agadès, mentioning Mr. Harouna, the mechanic in charge of Mr Guidoni's vehicles, who was asked to reserve a Land Rover for them. Bryce intended, when he got to Agadès, to ask Mr. Guidoni for an introduction to the local authorities at El Meki, and to enquire what he knew about the logistics arrangements the British were making there. In Agadès he was also to visit a Mr. Dambaba, Agent Special, to hint that Professor Ba would like to get his travel orders for a recent trip to Agadès stamped as soon as possible.

After saying a cordial goodbye to the Préfet they went to the US Embassy for a courtesy call. The Ambassador was busy with someone from South Vietnam, so they were shown round by Paul Inskeep. Bryce remarked that the Embassy seemed a pretty fancy place for a country whose only export to the US was an annual $26,000 worth of goatskin for gloves, and whose chief import from the US was $ 600,000 of used clothing supplied by Le Roi International and other enterprising outfits, which bought the clothing from US charities, for re-sale to the backward countries of the world. Bryce was scandalized and remarked that when the US taxpayer contributed to the Salvation Army or the Bishop's clothing appeal, he thought his cast-off garments were being given to the needy of the world. Cécile found out about this when she was asked by an American representative of Le Roi International to act as interpreter for him on a long-distance call from the Grand Hotel to a merchant in an outlying town, on whom he was trying to dump a load of clothing. The statistics quoted came from a brochure supplied by Paul Inskeep, and a telephone call to him later, confirmed every detail of Cécile's discovery. Although the Embassy building was pretty classy, Niger ranked as a hardship post—which meant extra pay. They borrowed some science-fiction paperbacks in the Embassy library, and had a lunch of hamburgers, French fries and cokes in the tiny lunch-room. The walk back to their hotel gave them a closer look at the town. It was fun seeing their shadows pointing southward at noon. The green of the vegetation was so bright it looked unreal. They were entranced by the big blue, white, and yellow, lizards, which scurried about and paused to do a few push-ups, before hurrying on. Every style of brightly colored and vivid clothing was to be seen, with some of the head-dresses "out of this world". Large ducks or geese flew over the Niger river, where kids swam in the water, and women scrubbed and pounded their laundry.

After resting in their room, they received a call from Morel, asking Bryce to bring up any mail for the NCAR group with him on Monday. He and Prantner were just about to depart for the airport to catch their plane to Agadès, so Bryce dashed down to show Prantner the telegram he had sent to Guidoni.

Bryce expressed regret that because of the strained relations, his illness and their dashing out on other errands, which may have seemed stand-offishness, they had not been able to get better acquainted with Prantner's group, and in particular, with Burritt. When Bryce said good-bye to him, Prantner said "I'll see you in Agadès". Bryce wondered whether he was coming down down the Arlit road with Burritt to meet him, or was he to come along to the Aïr, whether Bryce could get another Land Rover or not?

At this point Bryce inserts a few comments on Prantner, beginning with the fact that he could certainly get things done if he had enough money. He had already spent $ 45,000 on that year's site survey. He often seemed frustrated in West Africa by his complete lack of understanding of the spoken, and unspoken, language, which showed itself in a lack of consideration for the natives. He seemed to think that many of them were out to get him. He also seemed to resent Cécile's ability to make contacts and get things done, treating her as a rival rather than as a potential source of help. They were concerned at his lack of generous good-natured helpfulness, while acknowledging the fact that they had complicated his life.

They also felt that he was more concerned with the logistics rather than the success of observations, when he said that he was going there, "to prove that it couldn't be done", rather than, "I am going there to see if it is at all possible", and, on another occasion, when he said that he would only consider sites where he could bring in 20-ton trucks. Finally, Bryce noticed that Prantner used the term, 'seeing', in the sense of visibility, or of amount of scattered light, not in the more relevant astronomical sense, of quality of a point-image consequent on the passage of its light through a more or less turbulent atmosphere.

While this shows an appreciation of Prantner's position, in retrospect, one can only feel a lot of sympathy for him. He had come to a region of searing heat, lack of western-style facilities, all pervading dust, inhabited by people whose language he did not understand, and, when he tried to use the language they were supposed to understand, that he had presumably studied in the USA, they seemed to take a perverse delight in misunderstanding him. Then came these independent investigators—one can speculate, according to whatever his background may have been, that he counted their ivy-league eastern background against them, or alternatively, that they came from that upstart university in the south—these individuals, who had got him bad-mouthed with Ambassador Murphy, seemed to have a kind of black magic ability to hob-nob with all the most important people wherever they went, spoke the language and were given private hospitality, and who knows what they might say back in Washington.

On Friday June 23, the DeWitts were up before the crack of dawn for Cécile to catch the limousine to arrive at the airport by 5.30 a.m. She was still very tired and taking intestinal pills, but promised to seek medical help when she got to Paris.

Bryce did not go to the airport with her, and wrote in his diary that things were likely to change since he did not have her capacity to get things done rapidly. He wrote "She has dragged me by the hand half across Africa. But I would not have it any other way. Her presence has been indispensable . . . If I go on an eclipse expedition I shall want her there too."

Bryce had a restful day, only going down town to the Service des Mines, Ministère des Affaires Economiques du Commerce et de l'Industrie to see if he could get some good maps of the Aïr, as had been suggested to him in Paris in his telephone conversation with Profesor Trichet, and by the Préfet of Agadès. The staff were very cordial, but their maps were geological and did not show roads or mileages. They suggested going to the Service Topographique et du Cadastre, but the DeWitts had already been there, only to learn that they were supplied with maps directly from the Institut Géographique National in Paris. There were no usable maps of the Aïr at the desired scale of 1: 200,000. He also tried to locate Mr Mouren, a pharmacist, who owned the Hertz concession, suggested by Caldéron as a source of information about the Aïr, but he was out of town and nobody seemed to know when he would be back.

Bryce went to bed late and got up late the next morning, Saturday June 24. During the night he had woken up with a choking sensation to find the air of the room filled with an exceedingly fine dust, and looking out of the window, saw, what appeared to be a first-class dust storm, with visibility down to a few score yards, with the dust getting in through the air-conditioning filter system. He presumed that a great desert storm must have been raging up north to carry the dust down to Niamey. Later, it began to thunder and rain, which washed away the dust outside, but left everything in the room still covered.

His day was lazy, spent on writing up his diary, and, assuming he would never return to Niger, booking a trip with the Hertz people to drive up the river the next day to Ayorou, to see the hippopotamuses. He also wrote a letter of apology to Ambassador Murphy about Prantner's traveler's checks, and again tried out the Questar functions, still not fully operative.

On Sunday morning, he and his driver were off in a Land Rover at 7.00 a.m., the Peugeot 404 having broken down. Because of the overnight rain near Tillabery, the road was awash in many places, with water all over the plain and no culverts. It was nearly 11.00 a.m. before they got to Ayorou, but the long drive was worth it. There were numerous villages along the road with thatched huts and mud storehouses. Women were pounding millet with long poles, lots of birds with light blue feathers. There were no giraffe because these go into the bush in the rainy season, no longer needing to stay near the river for water.

Ayorou entranced Bryce. In its striking location on the river, it was market day, with many pirogues (hollowed-out tree-trunk canoes) ferrying people from one side of the river to the other. Hundreds of people of all kinds, in all degrees of dress and undress, with an astonishing variety of goods for sale, ranging from camels, goats, cattle and donkeys to meat and farm-produce and used clothing. There were seamstresses making repairs using hand-powered sewing-machines and blacksmiths repairing tools. Everybody was talking and swarming all over the market area, while others were loading goods onto or from pirogues. From the town, Bryce was taken upstream towards the frontier, turning off the road into the bush at the river bank, where they boarded a leaky pirogue manned by a couple of boys. About a mile downstream they spotted a couple of hippopotamuses. The boys stayed about 75 yards off, because the big bull didn't like the intrusion. He opened his mouth and emitted a rather threatening noise like a motor boat starting up. On the way back, Bryce asked the boys to put him ashore to walk back to the Land Rover. He passed a hut where five little boys ran out laughing, grabbed his hand and walked along beside him, examining him minutely, one of them running his fingers through the hair on Bryce's arm—in contrast to the smooth skin of local Africans. They turned very solemn when he tried to photograph them, and the same thing happened when they picked up a couple of boys in the Land Rover, when he asked them to stand in front of one of the termite hills which abounded in the area.

They returned to Ayorou, where Bryce ate a late lunch in a restaurant catering to Europeans, and then wandered through the market again, noting that infants on the ground in the hot sun seemed completely dazed.

They got back to Niamey a little after 6.00 p.m., after which Bryce had an early supper, packed, and went to bed.

Next day, he had to be up at 3.00 a.m. to catch his plane, a DC 3 with cargo-hatch, cargo roped in place on the port side with seats to starboard. Bryce remarked that it had been a long time since he last flew in one of these. They flew in smooth air at 4 or 5 thousand feet. It was dark when they took off, heading almost due east, and dawn showed visibility of 4 to 6 miles, with the horizon hidden in the haze. The country resembled that through which he had driven the day before, was absolutely parched for nine months of the year, but with the recent rains, had put on a faint patina of green. Bryce said that this vast region lying between the true desert to the north and the tropical forest to the south, was known in French as the Sahel, but he did not know the equivalent word in English. In fact, according to Larousse, the name was first applied to the littoral region of Algeria, where the population density was least.

Soon the plane reached the true desert on a northeast course, with visibility decreasing at first to 2 or 3 miles, then only one mile, before finally returning to 4 or 5 miles. The air grew bumpy as they neared Agadès, but ground level visibility increased as they came in for landing. The beginnings of the Aïr mountains could be seen to the north. Flying time was about 2 hours and 20 minutes. The plane carried ten passengers out of a possible capacity of 15. It took 45 minutes to get the baggage of about 5 suitcases off-loaded into a pick-up truck and driven 50 yards to the barrier surrounding the landing area. The passengers were not allowed to carry their own bags, and had to duck under the barrier which had no gate. As there were no taxis, anybody without his own means of transport was stranded, a category into which Bryce fell, but fortunately a nice Frenchman who worked for British Petroleum, gave him a ride to the Family House Hotel, the one suggested by the Préfet. This was pretty low by US standards, but there was air-conditioning and a refrigerator in each room with beer and soft drinks. An obvious effort was being made to divert people from the slightly larger and more centrally located (next to the mosque) Hotel de l'Aïr to the Family House.

After checking in, Bryce walked out with a young boy who served as guide, first, at least two miles, in a very hot sun, to Mr Guidonl's place way out on the edge of town, to see if he had a Land Rover for him, but Mr Guidoni had left town to go into the mountains, and nobody knew anything about the telegram that Bryce had sent. They then walked another three quarters of a mile to the office of the Agent Spécial, Mr Dambaba. This was mostly a courtesy call on behalf of Professor Ba. Mr Dambaba was very cordial. They talked of the eclipse preparations, and Bryce then mentioned Ba's Ordre de Mission (Travel orders), whose processing he would like to have expedited. Mr Dambaba then had his chauffeur drive Bryce to the military barracks on the edge of town, where he called on Captain Idrissa Harouna, again at Ba's suggestion. The Captain had traveled a lot in the region around the Aïr, and told Bryce something about the country, but nothing really valuable, such as a Land Rover, came of this visit.

Bryce then walked with his guide back to the hotel, arriving there about 10.30, to find Burritt and Morel waiting for him. They had driven from their camp on the eclipse line to get Burritt properly installed in the hotel. They had had a fairly rough time in the bush with dust storms and the heat, and the sky conditions had been terrible ever since their arrival. Already by 10.30 the horizon was invisible, the sky had no blue in it, and the Sun was a 3 or 4 finger Sun.

Morel and Burritt were both very friendly. Shortly after Bryce met them, Mr Harouna, Guidoni's mechanic-chauffeur arrived at the hotel to say that a Land Rover was indeed reserved for Bryce. This was the result of leaving a note at Guidoni's place saying where he was lodged. The only catch in this arrangement was that Prantner did not give Morel any money to pay for the vehicle, instructing him to tell Bryce to pay for it himself and to request reimbursement from La Count. This meant that Bryce's supply of money would be seriously depleted by the time he got to England. Fortunately, yesterday's travel expenses had gone on Bryce's Hertz credit card, but that would not do for Guidoni.

Just before noon, Bryce went on the roof of the hotel to make some seeing and haze measurements. The Sun was almost in the zenith, and to judge the seeing he only had one small sunspot near the solar limb. He made the seeing to be 7 or 8 seconds of arc, a situation which would make eclipse observations worthless. His two-color photometric observations were almost the same on the horizon and at an altitude of 45 degrees, but much less favorable than similar observations made earlier at Atar, although allowance had to be made for the fact that much of the light was being absorbed by the haze. This showed up clearly on Burritt's more refined apparatus, where he found the Sun's surface brightness to be less than on a hazy day in Baltimore, as it had been for the last four days.

At lunch Burritt and Morel reported the bad condition of the road from Niamey to Agadès, and that they had yet to see a usable sky for the eclipse.

However, they did emphasize that, with the early rains this year, sky conditions were completely different from those at eclipse time the previous year.

It was also clear that field conditions for eclipse teams—NSF was expecting about 150 people,—would be pretty grim at the Niger site. It was too hot to work between 11.00 and 4.00, and dust got into everything. Even with double-roofed aluminized tents, with walls that could be rolled up, things would not be comfortable. Local duststorms, better so described than sandstorms, could spring up without warning, with abrupt changes of wind direction, so that the windward side of a tent would suddenly become the leeward. With 150 people on a site like the one proposed, tempers would be bound to get short.

After lunch they all showered, with Burritt and Bryce going to bed, and Morel with his chauffeur to rejoin Prantner and Ecklund at their camp. During the afternoon a real duststorm came on and the Sun became invisible. Bryce awoke at about 5.00 p.m. and then went to help Burritt with a few errands, chief of which was to get Air Afrique to reschedule him on an earlier flight to Niamey. They then wandered round the town, had a look at the market, sat on the hotel roof until evening, at which time a lot of the dust cleared and was replaced by a scattered cloud layer. Then supper and bed.

A short time later there was a knock on Bryce's door to reveal a young black girl dressed in cheap western-style clothes. She offered herself for 2000 francs and said she was aged 14, which she may well have said on the assumption that men like girls young. What happened next was removed by Mikesell from the edited very frank version of Bryce's diary submitted to the eclipse committee. Let it be said that Bryce did not avail himself of the offer, and after a short conversation in the rather specialized French vocabulary appropriate to the occasion, sent her off with a present of 100 francs to hunt for bigger game.

On Tuesday June 27, Bryce was up at sunrise to photograph the Sun through the haze, and then had breakfast with Burritt. The driver arrived at 7.00 a.m. and they were soon on their way. The driver was a Targui, (i.e. a member of the Tuareg desert Berber people), named Sidi Ali, a superb driver like Mahmoud in Mauritania. He was said to know the Aïr region like the back of his hand and had many friends and relatives in the mountains. He normally shuttled back and forth between Agadès and Tamanrassé, using Harouna's Land Rover as a taxi for people crossing the desert. Bryce remarked that one often saw such vehicles, or more frequently trucks, packed with people. Bryce learned later that, had it not been for his telegram to Guidoni, which was treated as a request from the Préfet, Sidi would have been on the road to Tamanrasset that day. An excellent example of the value of making the right contacts. The road from Agadès to El Meki was, perhaps, a little better than that from Atar to Chinguetti in Mauritania, though it crossed several sandy dry river beds known as ouads or wadis, where a 404 Peugeot might get stuck. Incidentally, through the Arab occupation of southern Spain the word 'wadi' has survived in modern Spanish geographical names as 'guada', as in Guadaloupe, Guadarrama etc. Burritt said that this road was better than the direct one from Niamey to Agadès, and better than that from Agadès to Arlit, where Prantner's group's camp was located, over which heavy gasoline trucks of the semi-trailer variety ran regularly.

The road from Agadès to El Meki was terribly wash-boarded (corrugated). It ran through rather pretty country, not true desert. Regions of black volcanic rock were traversed by wadis bordered on either side by branching palm trees and other vegetation. Near these wadis they occasionally passed wells used to water a few date-palms and vegetables.

The sky on this day, as on the previous one, was dominated by haze. At 8.00 a.m. the Sun was a 3-fingered one, and the ground-level visibility was 4 to 5 miles. As the day wore on, this was reduced to 3 miles, which Bryce regretted very much, since it meant that many of the Aïr mountains all around them were quite invisible. With better visibility they might surely have seen many interesting distant peaks. The Aïr mountains are composed largely of very hard basalt, and rise abruptly with very steep sides, forming impressive cliffs in many places. Those which they could see loomed ghostly through the haze. Bryce likened the impression exactly to that of Mt Wilson seen from Pasadena through a Los Angeles smog. The haze was so white it seemed almost like a thin fog, but it could not be due to moisture because Eklund on the Arlit road had been getting humidity readings of between 7 % and 20 %. The haze puzzled Bryce for a long time, but decided it must be dust, though the conditions were totally unlike those that produced the haze in Mauritania. There was no choking dust near the ground, and the local breezes were both variable and not particularly strong. The haze seemed to be an inexplicable permanent phenomenon, not associated with the duststorms that Prantner's camp had experienced. The wadis were dry and packed with camel tracks so it was clear there had been no storm where they were, for many days. When he asked Sidi, or any of the other natives met en route where the haze came from, they all replied "Ça vient du Sahara", but the Sahara was there all the year round, and yet for many months the sky was clear. When he asked what happened when it rained, he was told that the sky was clear for perhaps a couple of days, and then the haze returned for 10, 15, 20 or more days. It was not until he read Burritt's copy of the British report on El Meki that he finally began to understand what was going on. According to the British report, the sky was relatively clear until the onset of the rains: then the haze developed. Actual rainstorms were too infrequent to lift a permanent dust layer into the sky, and, anyway, they would have a clearing effect. What evidently happens is this : There is a change of weather in the Sahara itself, associated with that northward movement of the Intertropical Convergence, which brings rain to the sub-Sahara. The Sahara at this time experiences some of its most severe lightning storms, (not to be confused with the great February sandstorms where sand is blown along by steady winds near the ground), which raise dust to high altitudes. It is the spill-over from these storms carried great distances by winds a few thousand feet above the ground that produces the haze.

By 9.00 a.m. the Sun had become a four-fingered one. At 10.00 Bryce took a sky brightness measure and noted a curious phenomenon. The sky was brighter in the red near the western horizon, (opposite the Sun), than it was 30 degrees above the horizon. The brightness peak around the Sun at the same time was very broad with readings increasing almost linearly with decreasing angle from the Sun until very close to it.

Bryce in fact, took few brightness readings on this trip, because the extremely poor sky conditions would make them irrelevant to those that the eclipse experiment would require, and in any case Burritt was taking readings of the brightness of the solar disk every hour, and would make his data available to the Texas group. Like Bryce's instrument, Burritt's had a red window and a blue window, which he aimed directly at the Sun, taking a simultaneous reading of its angular elevation, or rather, his instrument had a little scale which converted the angle into the equivalent mass of air through which the sunlight had passed, the unit being the zenith air mass, presumably at the particular altitude above sea level of the observer.

They arrived at the village of El Meki, a little before noon, and were surrounded by a circle of curious children, who were ordered back to a respectful distance by one of the men of the village. Bryce got out the Questar for a seeing observation. For what it was worth, this came out at 5 arc seconds, not too bad if the sky had been any good. As they still had a long way to go to Timia, Bryce did not tarry to let the children look through the telescope, but turned over a couple of kaleidoscopes to the man in charge, who following Bryce's explanation that these were for all the children, turned them over to two responsible boys in the crowd with an admonition to give everyone a turn.

They ate a small lunch from their concentrated food supplies in the shade of a small tree at some distance from the village, and then were off again. The road between El Meki and Timia was much worse than that between Agadès and El Meki, and could not, by any stretch of the imagination be regarded as traversable by any vehicle other than a Land Rover or its equivalent. It took over five hours to cover the 90 km to Timia.

The road wound over rough rocky ground, plunged down steep embankments, crossed innumerable wadis, climbed extremely steep grades, and in places followed the wadi bottoms themselves for several kilometers. There, a four-wheel drive was essential to avoid being trapped in the sand. The road ran between two mountain massifs, the Agalak massif to the north and the Baguezane massif to the south. The area between them was criss-crossed with wadis that flowed in and out of one another, and sometimes spread over regions a kilometer or more across. In this area one got a vivid impression of what happens when heavy rain falls in the mountains, with flash floods leaving one hopelessly stranded, at least for several days, until the sand dried up enough to become passable again. The nervousness inspired by this impression was not lessened later in the afternoon when heavy clouds built up over the mountains to the south. To this remark, Bryce appended a note: "This observation implies that slant visibility (i.e. into the sky) was greater than horizontal visibility at ground-level. It also implies that the densest haze was concentrated within a few thousand feet of the ground."

By 2.30 in the afternoon the air-temperature had reached 40°C (104°F), at which time the sky brightness was found to be remarkably uniform, outside of the peak near the Sun. The peak itself was both lower and narrower than earlier in the day.

At about 4.00 p.m., having passed two ostriches on the way, they arrived at a little village containing a small building used by the Service des Mines to house supplies. Here they stopped to get ready for the final assault, to let the engine cool, and check radiator water, while Sidi and the other driver palavered with the locals. The final assault consisted of an hilarious ride up the middle of a wadi, with Sidi constantly shifting gears as conditions required, until their way was blocked by a dry waterfall, about 40 or 50 feet high, which must be quite beautiful during a flood. The only bit of honest road between El Meki and Timia occurred there. A winding strip blasted out of the rock with a very steep pitch and smoothed with a stretch of concrete only 50 feet long, carried one to the top of the waterfall. Just before they reached this point, the incredible happened. The haze cleared, the sky turned blue and visibility increased to 20 miles. They were in a different world. They resumed their dash up the wadi and after two or three kilometers reached journey's end right out of a romantic novel. There, on a peak dominating the wadi, stood an abandoned French fort, with, at its feet, the date palms of Timia nestled along the borders of the wadi, green against the brown and black cliffs. They stopped at a tree standing alone in the wadi, and while Sidi exchanged the ritual of Arab greetings with the villagers, Bryce set up the Questar for a seeing measure before the Sun set behind the cliffs.

Burritt, with his photometer, got by far the brightest Sun reading that he had obtained during his whole stay in Niger. Bryce's result for the seeing was 7 seconds of arc, far from the 3 seconds at noon at Chinguetti, but tolerable considering the much increased length of atmosphere traversed by the low Sun's rays. He continued to have trouble with the Questar drive, even now with the voltage inverter hooked to the Land Rover battery.

The clearing of the sky put Bryce and Burritt in a quandary, since they could not honestly see trying to bring in a big eclipse convoy over that road but it did suggest that altitude did make a difference, provided that that was the cause of what they had seen, and not a general clearing over the whole area. They felt they could not determine surely until the next day. If for example the haze returned on the way back to El Meki they would know that the clearing was due to altitude. Bryce regretted that he had not brought his altimeter with him to Africa to determine just how far they had climbed. One of the maps seemed to show a height difference between El Meki and Timia of a little over 500 meters. He was not sure if that could be trusted, but he did think that they might well have climbed between 1600 and 1800 feet, and, from the general lay of the land, it could not have been much more than that. A small group of boys, 10 to 15 years old, gathered while he was using the telescope, but because there was only one small sunspot, and the Sun would soon be behind the cliffs, he did not set up a show immediately, but told them to come back after moonrise, and he would then show them lunar mountains. One of the boys asked if he would like to see some rock-gravures, created by prehistoric inhabitants of the Sahara, and he was then taken for a mile walk each way, to see a couple of very faint images, 24 inches high, scratched in the rock, showing a deer or a gazelle, plus what appeared to be writing in a strange alphabet. What he saw on the walk, which took them past the palm groves, was of more interest. Most of the men of the village were busy, some tending the cows, which pulled leather buckets of water for irrigation out of the wells, some repairing dykes in compartmentalized gardens, and some enlarging their plots. Everyone was courteous and friendly. Many of the dates were nearly ripe, and there were pomegranate trees also with nearly ripe fruit.

When he returned to the Land Rover beside the tree in the wadi, he spread his ground cover and sleeping bag out beside Burritt, who lay sprawled in the sand. He felt that Douglas was having a rough time, not having had much outdoor experience, and this being his first encounter with the medieval society of a poor country (Bryce said he spurned the euphemism 'underdeveloped'). On top of that, Burritt had a head cold and sore throat, probably brought on by the dryness of the air, as reported by the British at El Meki. He was feeling too miserable to get up off the ground when asked whether he would like to go see the gravures. Bryce felt that Burritt dreamt only of getting back to Baltimore, but he did get up, soon after Bryce returned, when the muezzin began chanting : "Allah, il Allah, il Allah" in a voice that could be heard all up and down the wadi. Some of the men assembled outside the low building which served as a mosque, while others prayed where they stood, or knelt with heads touching the ground, all facing eastward. Bryce and Burritt walked over for a closer—but respectful—look and then returned to a supper of powdered milk and Familia mixed with Agadès tap water.

Both men then flopped down on their sleeping bags and tried to get comfortable for the night, which, for Bryce, mainly meant pounding a depression in the sand for his hips. It was still too hot to get under any sort of cover, and sleep was impossible, because the two drivers, on their blankets nearby, and two men of the village, were engaged in a long discussion of the politics of Niger and of some of her neighbors, such as Algeria and Libya. Even so, after a day of cradling the Questar on in his lap over the roads they had traversed, Bryce was glad to lie down. It was soon completely dark, and Bryce began to wonder whether the boys of the village had forgotten his promise, but, before long, the eastern sky began to glow faintly, and at 9.00 p.m. the Moon poked her face over the eastern cliffs. At that moment the boys quietly appeared. The haze had completely disappeared and Bryce could orient his telescope correctly by reference to the pole star which was visible in spite of its very low altitude (nearly equal, of course, to the very low north latitude of their position). One by one the boys looked in the eye-piece, followed by the men, who had, for the moment, abandoned their politics. All seemed very pleased, but there were not so many questions as there had been in Chinguetti. Bryce thought, possibly because he had given out his fanciest kaleidoscope, one with a revolving wheel, on his return from the gravures to celebrate his attainment of the most remote and exotic of his African adventures. After that, came Jupiter, with even the men being tickled by the fact that that bright star was a blob with four tiny white dots strung out on one side of it. Then, good night, and darkness descended on the wadi.

Next day was Wednesday, June 28. The only creature Bryce encountered in the night was an ant crawling up a pants leg. By dawn the temperature was down to 20°C (68°F), causing Bryce during the night to don his hooded anorak and later to crawl into his sleeping bag.

He got up before dawn and saw two stars in the east, one brighter than the other, one of which, he was delighted to think, might be Mercury, which he had never seen before, but they turned out to be Venus, in a very thin crescent phase, and the other, Saturn, tiny compared with Jupiter and "way the hell on the other side of the Sun ". He began a series of zenith brightness photometric measures, and took the occasion to make a seeing observation, which, even with the very low Sun gave a value of not more than 3 arc seconds in the very quiet air. Bryce had been awakened before dawn by the call of the muezzin, which was repeated after sunrise. The only other sounds in the night had been the wail of an infant and the bray of a donkey, sounding like the horn of a diesel locomotive.

After a breakfast of powdered milk and Familia, Bryce and Burritt were off to an early start, the former being sorry not to climb up to the fort, but yielding to Burritt's desire to get back to civilization. The visibility was, initially, 15 to 20 miles, but by the time they got to the Service des Mines storehouse, it had dropped to less than 10 miles. Here the drivers again stopped for the usual palaver with the village folk. Sidi took the occasion to buy four chickens, which he crammed into a small cardboard box and covered with a tattered gunny sack, to take back to his wife. The chickens were reasonably quiet en route and squawked loudly only when the Land Rover suffered excessive jolts.

The visibility rapidly deteriorated as the altitude diminished and became as bad as it had been on the previous day, prompting the conclusion that altitude really did improve the sky conditions. On the return journey, the sky was, if anything, even worse than on the day before. Heavy clouds began to build up over the mountains and spread to the westward. By the time they reached El Meki, having passed a bounding group of five desert gazelles and two guinea-fowl on the wing, the sky was completely overcast.

There was a small rest house at El Meki, which was opened long enough for them to have a light lunch. All Bryce had was a concentrated bacon bar (1500 calories) and some Agadès water. During the drive there, the temperature rose to 41°C (106°F). On this stretch, they passed several Touaregs in deep blue boubous, with white turbans wrapped around their heads and faces, leaving only their eyes showing. They were mounted on camels with saddles having jaunty trident-shaped pommels. Each man sat proudly erect, holding firmly on to his large sword. By the time they reached Agadès in mid-afternoon they had left the clouds behind, but the haze remained as always. Before saying goodbye to the drivers Bryce gave each of them one of his two remaining kaleidoscopes to give to their children, for he had learned that they both had families. He proposed to give Sidi his carton of cigarettes.

After showering at the Family House, Bryce spent the rest of the afternoon resting, but got up just before sunset to take some brightness readings, which he curtailed as the Sun began disappearing in the haze. He photographed it to show its dimness, supped with Burritt and went early to bed.

Thursday June 29 was spent working on the journal and killing time until the Niamey plane next day. He photographed the rising Sun through the usual haze and, after breakfast, walked around the town with Burritt to take some photographs. It was already hot walking round the sun-drenched town, but when they climbed to the top of the 80-foot minaret (made of banco) for which Agadès is famous, they found a steady cool breeze. Bryce remarked that, had there been room to lie down, he would have been happy to stay there all day.(Figure 32).

Back at the hotel he found Mr Harouna with the bill for the Land Rover. The cost was remarkably low, only $ 140 for the two days, much less than Hertz was charging Burritt. Bryce was charged a flat rate per day, driver included, plus gasoline, with no kilometrage. Bryce was sure this courtesy was extended on account of the Préfet. He did not meet Mr Guidoni, but Mr Harouna was an exceedingly nice man.

Bryce stayed up late writing in his journal. He would have spent more time on it during the day, but Burritt was going stir-crazy and wanted someone to talk to. Also, the hotel proprietor, Mr Boudon, was around a lot, and he was a great bavard (chatterer). He told Bryce all about his life in the French army in Morocco and West Africa in the thirties (Boudon was 60 years old), about exploring a pass in the Aouker escarpment in a model A Ford, driving the vehicle across the desert to Marrakech, and of being captured by 'les Maures'. He also told him the story, which Bryce already knew, via Cécile from her father, of the rise to fame of André Citroën through his development, for his general,of a method of mass-production of artillery shells during the first world war. Bryce found Boudon to be well informed and accurate on scientific topics, who responded intelligently to questions about Saharan weather. By this time, Bryce had read Burritt's copy of the British

Figure 32: The minaret at Agadès.
(Photo. Dr. John James).

report on El Meki, and was beginning to get a clearer picture of the weather situation in the region.

Boudon was apparently married to an Arab woman, whom they never saw and to whom Boudon never referred, but they did see his very sweet daughter, aged about 13 or 14, at nearly every meal. Boudon himself was thought not so sweet, still a bit colonialist, who, however, served a good cuisine in a thoroughly French style, right down to puddings for dessert. In the evening they settled accounts, and packed their bags ready for an early departure.

On Friday June 30, Burritt, who had an alarm clock, wakened Bryce at 6.00 a.m. Breakfast was ready a quarter of an hour later, and five minutes after that, Mr Boudon arrived to take them to the airport. They were the first passengers there, and while waiting for the plane to arrive from Arlit, Boudon took them to see the airport meteorologist, whom Bryce plied with questions about Saharan weather. He generally confirmed the picture that had been forming in Bryce's mind, but added two points of special interest: (1) The dust of which the current haze was composed is concentrated relatively near the ground, most of it below 1500 to 1800 meters (Bryce:—very like Los Angeles smog!) (2) The arrival of the haze in the Agadès region depended, not only on the occurrence of electrical storms in the Sahara, but also on the presence of winds to carry the dust in the right direction. A change of wind direction could sometimes even clear the sky during the rainy season. He agreed that the haze had arrived earlier that year than usual, but when asked what he thought the chances were of its arrival before the thirtieth of June in an arbitrary year, he ducked the question and referred Bryce to the files of the meteorological service in Niamey.

The DC 3 from Arlit arrived a bit late and was jammed with people on both sides, with no port-side cargo this time. The crew seemed to have no record of their reservations and there were no airline personnel in the airport building. It was beginning to look grim, with the pilot saying "Il-n'ya plus de places", but he suddenly relented and allowed Bryce, Burritt and one other new passenger to climb aboard with their luggage. The steward had given up his seat and some cargo had been shifted around, so that, miraculously it seemed, every passenger had a seat. Evidently on Air Niger, reservations were given automatically on the assumption that there always would be room on a DC 3. No one ever really bothered to check. Bryce noted,—underlined,—*Eclipse teams please take note.* Bryce felt that the plane was pretty close to being overloaded, but the whole procedure was repeated at Taoua, the only other stop en route. Here, even more passengers were squeezed in, to Bryce's mystification. They arrived at Niamey shortly before noon and were met by the Grand Hotel limousine. Burritt, anxious to get out of Africa as quickly as possible, wanted to stop at the American Embassy, where he had left his return airplane tickets, and then go straight to the Air Afrique office to see if Mrs Buresi could get him on the same flight to Paris that Bryce was taking. So to the Embassy on the other side of town they went, where they found Paul Inskeep, who told him that his tickets were at the SABENA office. As it was then too close to siesta time to allow them to go both to the SABENA and Air Afrique offices, Burritt suggested that they should send the driver to the hotel with their luggage, and stay for lunch at the Embassy, which they duly did. After that, the Embassy chauffeur took them to their hotel, where Bryce wrote in his journal. Fortunately he had already reconfirmed his Paris flight at the Niamey airport, but he phoned to double check it with Mme Buresi at the Air Afrique office.

At 4.00 p.m. they set out for the SABENA office where they picked up Burritt's tickets and went on to Air Afrique. There Mme Buresi was her usual efficient self. Bryce was properly confirmed,but Burritt was only on standby status. She would keep his tickets until 11.00 a.m. the next day, but, in the meantime arranged things so that his Paris-Washington flight could be confirmed as soon as the Paris-Niamey confirmation came through—even after 11.00 a.m., when he should proceed to the airport as a standby passenger, so that Air Afrique—actually in his case UTA—would automatically take care of his transportation from Le Bourget to Orly and his overnight accommodation in Paris. Then to Banque Internationale pour l'Afrique to help him change some US dollars into CFA francs and then to the Musée National du Niger, where he wanted to buy some gifts for his family in the museum shop. This was a nice ethnographic museum consisting of several buildings, a model village, and a craftsmen's workshop area. There was also a small zoo with crocodiles, giant turtles, two hippopotamuses, three lions, a hyena and a cheetah, some monkeys and a few other smaller mammals, an African elephant, and a nice aviary with eagles, falcons, vultures, a secretary bird, and a number of comical big birds of three or four totally different species unknown to Bryce. He left Burritt at the gift shop and went back to the hotel for supper at 7.30, after arranging to meet for breakfast in the morning. His African adventures were nearly over.

On Saturday July 1, they breakfasted together beside the swimming pool. The Hertz people arrived at 9.00 a.m. and helped him leave final instructions with them for the loaded Land Rover, which was due to arrive back in Niamey in a couple of days. Then to Air Afrique, where Mme Buresi had Burritt's flights all the way back to Washington confirmed, and notification sent to have Air Afrique look after him in Paris. At the hotel, they managed to persuade the staff to give them lunch early, and, contrary to usual custom, put all the courses on the table at the same time. They then left to catch their 1.15 p.m. flight to Paris. The flight, originating in Ouagadougou, arrived soon after they did, while they were checking in their luggage. The staff were very strict about weighing it, since this was the only flight of Bryce's entire trip from Austin that was fully booked. They charged $ 20 for about 4 kilograms, not including the Questar which he always carried aboard and put under his seat. Among the passengers who had disembarked from the plane to stretch themselves was Dr Kogan of the Peace Corps, who was returning to the States. Bryce learned that he was a general practitioner from Hollywood, California.

On boarding the plane the stewardess at the top of the steps called Bryce by name, and asked if arrangements had been made for his bill at the Grand Hotel. Evidently Air Afrique was supposed to foot the bill for his expenses since arriving back in Niamey, because the flight to Paris was the first connecting one from Agadès. This was the first he had heard about it. The stewardess was the one with whom he had reconfirmed his flight the day before, and she had not mentioned it to him, nor had Mrs Buresi. It was too late to do anything about it, and Bryce only mentioned it as a useful item of information for future travelers in the region.

The aircraft, a DC 8, was completely full, with not a single empty seat. With no ground-based air-conditioning at Niamey, the cabin became unbearably hot for half an hour, and even after take-off, the blower system was not very efficient, so that things only became comfortable when they reached cruising altitude.

Because the plane was so full, Bryce did not have a window-seat, and could see little of the surroundings. There were no signs of storms anywhere and the haze reached 20,000 feet and seemed uniform, of a brownish or yellowish white color. This lasted all the way to the Algerian coast, where it disappeared, and was replaced over the French coast by low-lying European cloud cover. There was little of the local variation in seeing conditions that he had noticed over the Atlantic coast deserts, and when he could see the ground through a steep enough angle, he estimated the visibility to have been about 10 miles.

They landed at Bordeaux about 5.30 p.m. to go through customs, where everybody began to shiver in the cool wind. After 45 minutes they took off again, now with empty seats, with an hour to Le Bourget, another hour to reclaim baggage, and after saying good-bye to Burritt at the Air Afrique counter and ensuring that he would be properly looked after, Bryce took a taxi, which in the heavy Saturday night traffic took another hour, and it was after 10.00 p.m. when he finally reached rue Madames.

# Conclusion

Bryce ended his journal with a summary of conclusions which we reproduce here:

The strongest single feeling I have acquired from the trip is apprehension. I am scared of the Sahara's potential for completely ruining any chances of taking useful photographs on eclipse day. Even were the Intertropical Convergence to move northward at its statistical average rate next year, instead of jumping the gun by a month, June 30 is so close to the average transition date in the Agadès region that it gives me the shakes. Granted, no two years are alike, but the desert is perfectly capable of presenting some other diabolical aspect of itself at that season next year . . . and there goes $ 250,000.

I am scared also by the philosophy of the American logistics planners. In talking to the various groups that hope to observe the eclipse, they have a common denominator. Each group wants to be as close to the eclipse center line as possible. Therefore, the logisticians have gone all out to find exactly where that line is located—(with sextants, theodolites and all), and to set up the American camp there, be it in ever so hellish a location. Furthermore, the NSF have advertised to all American universities the facilities that it plans to provide at the camp, and has invited them to submit proposals. There can only be one outcome : Groups that normally, would not have considered making the effort, will send teams to the camps, composed of scientists who have neither the experience, nor the mental outlook required for an expedition, and who imagine that, with basic amenities provided, they will be able to devote nine-tenths of their time to purely scientific matters. They will be deceived. When they find themselves gasping for breath under the hot tents or struggling with the tent flaps in a duststorm, or when they find that they can't go to town to fetch an expected piece of equipment because someone else has priority on the truck, tempers will explode.

Consider the impression the American effort will make on others. It will be totally unlike the other national groups in its sheer size, in its approach to logistical problems and, above all, in its basic outlook. Its aim will be to provide an American environment in the desert, largely insulated from the surrounding country and populations, like SOMIMA in Mauritania. It will appear to others to be just what it in fact is:—an expedition mounted like a military operation, backed by lots of money—and the analogy to logistics operations in Vietnam will not go unremarked.

I say this not to condemn the American approach. Some of this sort of thing is inevitable. However, I would like to make a strong appeal for greater flexibility on the part of NSF, if it is not already too late. For example, instead of pitching camp in the middle of nowhere on the Arlit road in Niger, why not find a wadi near a well up the El Meki road, as the British plan to do, even if it is 5 or 10 km off the eclipse center line. A wadi with its clean sand, bordering palms and surrounding greenery is infinitely more pleasant than the dirty desert to the west. Let those who must, at all costs, be on the center line make their camps a few miles off. Could the main camp not be used as a base-camp for satellite groups? Could some of the satellite camps not be as far away as 50 miles, provided they are willing to supply their own generators for electric power?

To my mind the ideal, in the beginning, would have been for the NSF to indicate that it would only support those groups that are capable of mounting expeditions *entirely on their own,* and then to step in and smooth the way for these groups by helping with their logistics. In this way the NSF would have had to deal only with persons who thoroughly understand the problems of operating in the field, who are able to cooperate effectively in return for NSF's helpful hand. It is too late for that now, but perhaps some of it can be salvaged.

I recommend most strongly that one of our sites be the airstrip at Chinguetti, (Figure 33),provided the NSF is willing to transport our equipment thither. Could the NSF not charter a DC3 in Dakar, which would make round trips from there, ferrying at least the more delicate equipment to the groups at Akjoujt, Atar and Chinguetti? No group has yet officially requested the use of Chinguetti, either of the NSF or of the Mauritanian government. Therefore, it is important for us to do so *as soon as possible.*

Because of the extreme importance to our observations of having the minimum amount of dust in the atmosphere, I feel that we must at all costs go to Chinguetti if we go to Mauritania at all.

By the same token, I feel that we should give serious thought to trying to reach Timia in Niger. It is clearly a much more difficult location than Chinguetti. It cannot be considered if it proves to be impossible to pack our delicate equipment carefully enough to withstand the jolts of the road or to assemble all of our equipment from components, none of which must not weigh appreciably more than 500 kg. It also cannot be considered if we do not have access to Land Rovers, and to drivers who are thoroughly familiar with the area. In this connection, the importance of having access to funds (e.g. NATO or private) that are not subject to the "buy American" restriction emerges. If we could buy just two Land Rovers, (for which parts and repairs can be obtained in Agadès), we could consider Timia. If the NSF were to set up its camp on the El Meki road, we would bear only the expense of transporting our equipment and supplies from there. If the NSF chooses the Arlit road it would be better for us to haul it directly from the Agadès airport.

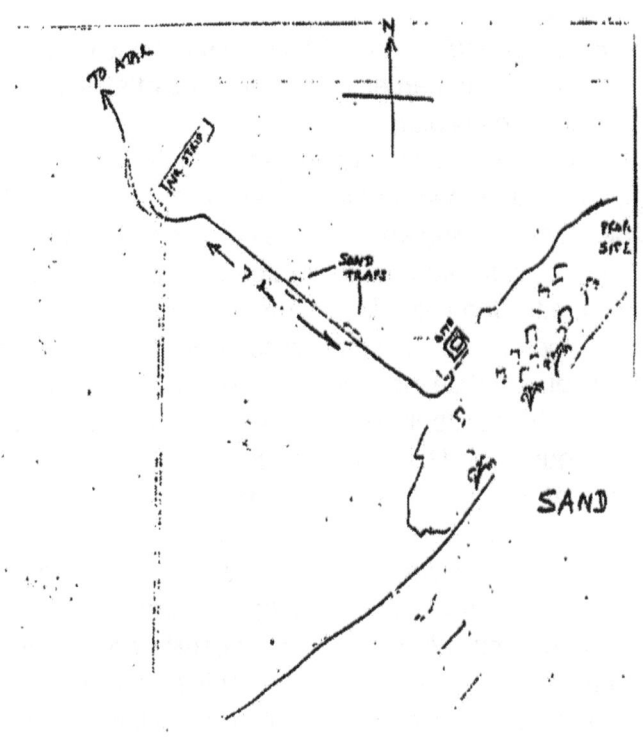

Figure 33: Bryce's sketch of his recommended
eclipse observation site at Chinguetti.

Of course, if access to large refrigerators, helium or a fully stocked darkroom is essential to our operations, then we must decide to set up shop at the main camp and trust that weather statistics will bring us favorable skies next year. My only reason for recommending Timia at all, aside from its charm, is that even modest increases in altitude *do* make a difference.

Bryce added a copy of the British site inspection and a careful list of the names and addresses of all the individuals mentioned in his diary.

This conclusion was a rather glum document, criticizing the way things seemed to be getting organized back in the USA, and worrying about access to sites. He does however, come out with high marks in some respects for Timia and Chinguetti. Some of his recommendations on desert travel seem open to criticism, in particular his feeling that camping in a wadi would be more comfortable than in the open desert. Perhaps he had, only for a moment, forgotten that a wadi is a watercourse, and that all desert features are sculpted by rare floods of water. These can be produced by heavy rain, so remote from a given location that it goes unremarked until the wadi suddenly becomes a torrent, as has happened in many a case of human disasters. Apart from that, Bryce was obviously ambivalent about the deserts and semi-deserts he had seen, half-fascinating in their sights and splendors, half-threatening in their potential for involving humans in fatal circumstances.

## Second Opinions

Bryce's report was submitted to the committee in Austin after some editing by Alfred Mikesell. The version from which we have quoted so extensively is the uncut text.

The British site-inspection document came from Dr. J.P.James of the University of Manchester. After a week of overland desert travel from Tangier to Niamey, he had arrived at Agadès on March 17, where he was looked after by the British mission to the Republic of Niger and given the use of some of their Land Rovers. He had identified two places where the eclipse center-line crossed a desert track, one of which he deemed quite unsuitable, while the other would be quite feasible if the expedition were properly planned. (Figures 34, 35)

He then went on to describe the countryside as rough rather than sandy, with a sparse population of the desert tribes. These people cultivated an acre or two of garden, irrigated by well water. (Figures 36,37). The inhabitants were poor, kindly, and polite, people, speaking some French, but generally only reading and writing a desert language. There was a track to the village of El Meki, about 60 miles north of Agadès, and a further track to a wadi, which meandered for miles north of the village. The center-line passed very close to this point, and it was there that he recommended a camp be established. The wadi was 30-50 yards wide, lined with palm trees and other greenery, and flanked by flat sandy ground with ample space for tents and apparatus. A well could be dug by local labor, also available for housekeeping, readily obtainable in El Meki or Agadès. The ground was covered with sand to a depth of a couple of inches, overlying a firm subsoil, but he recommended that concrete rafts be made to support apparatus in case of subsidence after setting up. Tents could easily be pitched and locally made matting made an excellent floor covering.

The other site examined was where the Bilma track crossed the center-line about 130 miles east of Agadès, near a large solitary three-thousand foot basalt mountain, known on the Michelin map as Adrar Azzaougar, but to the local Tuareg as Im-zigguir. There the ground was sand, quartz stone, and very sparse scrub grass. It was gently undulating with an occasional tor of white quartz, and very close to the edge of the Great Sand Sea,(*Grand Erg du Ténéré*). There was no shade at all, nor any ground flat enough for a single

Figure 34: Agadès seen from the minaret.
(Photo. Dr. John James).

Figure 35: The market at Agadès.
(Photo. Dr. John James).

Figure 36: The garden at El Meki. Saharan oasis waters generally come from deep, very ancient, aquifers.

Figure 37: At El Meki Touareg neighbors came calling. (Photos. Dr. John James).

camp. Chiefly, though, it could be reckoned untenable because of the heat, not to mention the monotony of the scene—this in mid-March. In July, with the higher temperatures, it would be an extremely unpleasant place to spend four or five weeks. (Figures 38,39)

James had looked at the meteorological records for this part of the Sahara, and he too, had consulted the meteorological officer at the Agadès airport. Most of that information dealt only with the general region, and he put more faith in the informal talks he had had with the people in El Meki. They were quite positive that at the end of June—a few weeks before the rainy season began—the sky would be quite clear at El Meki.

The climate generally was healthy and disease uncommon. James sampled—in the cause of science—the well water from various parts. It occasionally contained larvae, but it never gave him any intestinal qualms. He had with him a supply of sulfaguanidine tablets, but the bottle remained unopened. Local fresh meat of adequate quality was available. He had eaten and enjoyed, beef, mutton, gazelle, goat, and something that might have been camel. Vegetables were for sale from people living by the wells. Tinned food was expensive—it came in by air-freight—and James had opened negotiations with an Arab merchant in Agadès for bulk-buying of tinned food. Surprisingly, in an Islamic country, beer and spirits were available locally and were not particularly expensive.

Surface transport was by Land Rover or camel. The latter could carry about 200 lbs about 25 miles per day. (Figures 40,41). Land Rovers averaged 4 km per liter of gasoline, and the average speed on the rocky desert tracks was about 30-40 km per hour, though on the hard flat sand of the Ténéré the speed was much higher. It was imperative to carry water, even on the short trips from Agadès to El Meki, in case of breakdowns. Five liters per person would be adequate on this frequented road. Normally one allowed 8-10 liters per day for general consumption. There were many Land Rovers in Agadès, but it would be unwise to assume that it would be possible to hire one as needed. There was a good mechanic in town who was familiar with this type of vehicle and kept a supply of spares.

James identified three major problems. The first was the extreme heat and light. He remarked that dark glasses were not necessary, but that adequately-sized lenses and rims were needed to avoid conjunctivitis. He quoted temperatures in March as 90°F at 08.00 hours, 105°F at 12.00, this being maintained until about 16.30, after which it gradually declined. At one village in the Aïr where he spent some nights, it was 81°F at midnight, but on the mountain at his second site it was only 52°F by four in the morning, and he needed blankets. He was told that at El Meki the Sun would be hotter in June and "slightly" more uncomfortable. In March it was not possible to do any serious work between 11.00 and 16.00, and the rule was to take a siesta after a light lunch at 11.00. He said "I must emphasize the heat. It was much worse than I am accustomed to in tropical countries".

The camping need was thus for shade rather than shelter. Metal parts left in the open reached a surface temperature of round about 100°C and ought to be painted with a thick white gloss enamel if they were to be cool enough to handle (e.g. brass knobs on coelostats).

Dress should cover the person to provide thermal insulation. Swimsuits were not to be recommended, and clothing should be designed to trap a layer of cool air next to the body.

The second problem was humidity, or rather lack of it, at less than 5 per cent. This made new arrivals immediately develop symptoms of a cold in the head, through dehydration of the nasal passages and larynx, and causing hoarseness of the voice. On the other hand, laundry on the line dried in 5-7 minutes.

Figure 38: On the map, a possible eclipse
observation site, rejected on closer inspection.
(Photo. Dr. John James).

Figure 39: Another reject. The famous single tree,
the only vegetation in the huge Teneré sand sea,
is visited by Dr. James. (Photo. Dr. John James).

Figure 40: One of the famous salt-caravans
is glimpsed in the distance.
(Photo. Dr. John James).

Figure 41: The caravan passes close by on its way
to a market for the salt carried by the camels.

Dust could be a problem. Occasionally there came a strong wind producing a duststorm which might continue for two days. This phenomenon was not to be confused with a sandstorm, but could still be an embarrassment. The particles, of which he took home a sample, were not particularly abrasive, and under the microscope were rounded and smooth like pebbles : but they would get into every crack and gear-tooth making a disagreeable paste when mixed with lubricating oil. He recommended large polythene bags for protection against dust, but cautioned that there would be a very pronounced 'greenhouse' effect and the whole package ought to be shielded from direct sunlight. He intended to investigate the purchase of silvered balloon fabric for this. One of the camp tents should have sides to let down in a duststorm. He was unable to get any information on the frequency of these storms but guessed at not more than one a month.

Apart from all that, camping seemed to be pretty normal for the tropics. Fly tents were preferable and it would be possible, and even desirable, to sleep on camp beds in the open, with no more covering than a mosquito net, against flies, not mosquitoes.

He went on to discuss sanitary and washing facilities, the provision of refrigerators and so on. Agadès was two and a half hours away, with medical facilities in a clinic with four French doctors in attendance. There was a reasonable air service from London.

James continued by sketching the rather ambitious plans for the expedition, including, possibly, the services of the RAF to fly everybody with their gear to Agadès, or a large charter aircraft for the same purpose, or to air-freight the gear, and to ship the Land Rovers to Cotonou in Dahomey, driving them to Agadès to pick up the air-freight there. He contemplated the arrival of personnel in three groups, starting six weeks before-hand (two people), six people five weeks early, and seven people three weeks early, and a liquid helium courier three days before eclipse day.

A later exchange of letters between James and Harlan Smith confirmed these general features of the Manchester proposal. James emphasised that the logistics were difficult, and great care had to be taken in preparing an expedition, since mistakes could literally be fatal. On a recent German 'safari' type expedition, one member had died while en route to Bilma (farther up the mountain chain from Agadès). This reinforced an unconfirmed story of an eclipse expedition in Mauritania trying to cross overland, with very good desert equipment, from the northern port of Nouadhibou, (formerly Port Etienne), to the eclipse line, which got stuck and had to be rescued by the Mauritanian military.

James also said that expeditions should be as small as possible, and that the local authorities were alarmed at the possibility of a hundred or so astronomers descending upon them and overstretching local resources, especially the water supply from local wells. James remarked that if they took an expedition to the region, they would try to limit it to twelve people.

There was no obvious difference between the mountains and the desert so far as dust in the sky was concerned (in contrast to Bryce's findings) nor, so far as he could tell, in the frequency of duststorms. The reason for choosing the mountains was one of temperature: "The ambient in the mountains is about 120°F at that time of year, and in the sand it can be 140°F. The record, near Tamanrasset, was, he was told, 156°F". When James was there, the sky in the mountains was not particularly clear, nowhere near as clear as in Mexico at the 1970 eclipse, but, with the Moon's shadow being bigger than usual they might expect a darker sky. He himself was planning to look at Doppler shifts in the lines of the F-corona, and was hoping for maximum darkness. He said he was firmly promised a clear sky at El Meki for June 30, and could well believe it, the sky being cloudless until a little cumulus appeared at three or four in the afternoon. To sum up : it was probably worth trying it at El Meki, but not the place for Texas' main effort, there being no advantage to the Texas program except a longer totality.

He had found it hot beyond belief, much hotter than anything he had experienced in the tropics before. There are no candles in Agadès, because candle-wax is a liquid in those parts.

On June 27, Al Mikesell had a long conversation with Russell Ulrich, Chief Meteorologist of the Pan American World Airways System, which included, among other things, reference to records of summer rainfall at a number of West African stations, including Gao, Tessalit, Wau and Agadès. The results for the last-named were, June : Nine-year average total 0.3 inches on one day : July same, 1.9 inches on 5 days, and August total 3.7 inches. It was remarked that Kenya had a bi-modal rainy season, April-May and November, as the Intertropical Convergence moved back and forth across that equatorial region, so that Lodvar got on average 0.1 inches on one day in June, and a total of 0.5 inches on three days in July. It was not cloudy between May and November.

Elsewhere, Ulrich said, the ITC bows north over central Africa, and does not get to Nouakchott even in July, nor to Akjoujt, even less and later. For Akjoujt no rain or cloudiness was noted in June-July. The northern limit of the annual migration of the ITC was only a little north of Dakar. Ulrich proposed retaining a meteorological consultant, and suggested going to that department at Texas A and M, which was so good that they should be able to field the essential question "What are the chances of clear weather at site?". The suggestion of a meteorological member of the team is less beguiling in practice than in theory. The best site is picked out by meteorological studies in advance, but once that has been done, nobody, except the news media, cares what the short term prediction may be. The site cannot be changed:

the observations will be attempted according to program no matter what: and what the conditions were like at eclipse time will be a matter of record, whether the program was a success or not.

Evans, the only team member with extensive experience in Africa, including some field work, was still in South Africa and not privy to the discussions in Austin, but wrote on August 8, evidently having seen the James letter, as follows:

"The more I think about things the less struck I am on the Central Sahara particularly having two sites there. I do not think I personally am young enough (he was 56) or tough enough to be a good risk as a team leader on the spot. I am not convinced, even in spite of Bryce's experience, that going up mountains is a good thing. The Mauritanian experiences may be valid, but there seems to be far too much of this business of side-deals which may land us in all sorts of trouble. As the French say, 'C'est magnifique, mais ce n'est pas la guerre'. It would be downright silly to ignore experience on the ground. If one can't work between 11.00 and 4.00 p.m. in the Central Sahara we shouldn't try to observe an eclipse there".

Referring to S.P. Jackson's 'Climatological Atlas of Africa', Evans said that this showed a minimum and low rainfall for a few stations immediately east and south of Lake Rudolph in Kenya. To the west there was higher rainfall, but the data were not new and there were fewer stations. Plateau heights of 5000 feet seemed attainable in the area, and there was a 9000 foot mountain near the eclipse track, though he did not have a detailed map with him. He remarked that Land Rover safaris to Lake Rudolph were run from Nairobi. "I would have thought a poor man's expedition would have a good chance if you think differential refraction controllable. This is the sort of place I wouldn't mind going myself".

Of course, this area seemed much more attractive to him as being almost identical in character with the kind of terrain of which he had extensive experience in other parts of Africa.

Among the documents dealing with this topic, there is a remark that, although the Sun would be much lower in the afternoon in Kenya, it was no worse than at the famous Lick observatory expedition in Australia in the twenties.(see Figure 13).

At some point a communication was received from Dr Rigutti, an Italian from the Capodimonte observatory near Naples, who described himself as the International Astronomical Union coordinator for the eclipse, saying that they had lighted on Timia, and inviting the Texans to join in there. However, by October 13 Mikesell was writing to James and to Rigutti with letters of thanks for their information and invitations to joint sites, and sending the reports of the reconnaissance by the DeWitts. To Rigutti Mikesell said, "Our only expectation is the location of a modest field party with one telescope at the NSF site in Mauritania".

# Choice

So the die was finally cast, it would appear from other correspondence to be mentioned below, that the scale of operations had been reduced from earlier ambitions by considerations that there would be little gain in trying to mount more than one expedition, that financial prospects were not as ample as originally hoped, and that there were serious problems about the provision of even one astrometric lens, let alone three. As the proposals took final form, the official US-backed logistics took over some of the planning, avoiding most of Bryce's fears and bringing to bear the almost military precision of organization the enterprise warranted.

We can jump a little ahead by noticing that on December 26 1972, the National Center for Atmospheric Research Facilities Laboratory in Boulder, Colorado, issued 1972 Solar Eclipse Communication No 1 over the signature of George Wm. (Bill) Curtis, Head and Scientific Coordinator. This was an extremely business-like document, detailing the NCAR management group consisting of Curtis, Prantner, Nelder Medrud, and Mrs Jane Aschenbrenner, the Secretary, through whom all correspondence was to be routed. Subjects which would be covered in due course would include transport of equipment and people, important dates, passports, visas, vaccinations, and accommodations en route and at the sites. There would be a medical doctor in residence at each site. One was to be at Chinguetti, where the heat and humidity were cited, and at Loiyengalani, a village in Kenya near the shores of Lake Rudolph.

Personnel were encouraged to get medical check-ups with their own doctors and a prescribed list of inoculations. Questionnaire forms for personnel particulars and applications forms for visas to Senegal, Mauritania, and Kenya were enclosed, one for each expected participant.

The whole expedition, including all approved American teams which had applied to the NSF, was described as the National Science Foundation 1973 Solar Expedition to Africa, which had selected the sites in West and East Africa in the Fall of 1972, and nominated qualified groups to receive logistics support, of which the Texas team was one.

## Meanwhile Back in the USA

To fix on a site was one thing, to decide what to do there and to provide the necessary equipment, were quite others. The roots of all this, covering finances, policy, program, techniques and instrument construction lie considerably earlier, and must now be examined.

At the very beginning Harlan had not been very enthusiastic, and Bryce recollects his saying that he wasn't interested in eclipse chasers. He, of course, came round quite early on, but Bryce was already in touch with old Princeton friends who had opinions and offered advice. These included

David Wilkinson, Phillipe Crane and the most senior, Robert Dicke, himself a theoretician of relativity, and the co-author of a system variant from that of Einstein. Because of his official status as a judge of other people's projects he could not take an active part in promoting the interests of any particular group, but did offer highly competent sophisticated technical advice on the problems of astrographic photography.

In plain terms the task was to take a photograph of the eclipsed Sun, conveniently hidden by the Moon, which would show the positions of neighboring stars in such a way as to allow these positions to be compared with those which would have obtained if the Sun were safely out of the way, and allow the differences of these positions to be measured.

The old cliché says that the camera cannot lie, but it certainly transforms the truth and introduces sophisticated problems in the interpretation of photographic records, particularly in a case such as this where very small displacements of star-images were to be measured with great accuracy. But before we even get to the photographic part of the problem we find that there are other sources of systematic distortion in the picture of the sky presented to us by nature and our record systems.

The first of these is the Earth's atmosphere, which, in its cruder manifestations can reduce the intensity and definition of the picture by cloud or dust absorption and scattering, and, even when these effects are absent, can introduce a turbulent medium between the source and the observer, which turns what might have been a perfect point-image of a star into a more or less fuzzy patch, with a greater or less central concentration of brightness. These sources have to be tolerated at the time of an eclipse, but, as we have seen, observers try to minimize them as much as possible by selecting sites with clear skies and calm air. Beyond this is the fact that the atmosphere of the Earth acts like a prism on light rays passing through it, both bending their directions and producing a slight spectral dispersion. The deviation effect has long been known and, even in the seventeenth century, was intensively discussed and ways to evaluate it were sought. When a star is seen at some elevation in the sky its light rays have impinged at an angle on the atmosphere of the Earth, which has acted like a prism and deviated them to a slightly more vertical direction. The result is that the star is seen by the observer at a somewhat greater altitude than would have obtained had there been no atmosphere. The lower the altitude of the star the greater the effect. It is small or absent near the zenith, but for objects seen low on the horizon the effect is very considerable. It is so large right on the horizon that when an observer sees the setting Sun just touching it, he would, if the atmosphere were suddenly removed, become aware that the Sun had actually set completely. So the prudent eclipse observer goes to some place where the event will take place well up in the sky—for example, at Chinguetti at an altitude of 61°. Since the effect depends on the amount of atmosphere through which the starlight has passed, there is also an advantage in observing from a site as high as possible, with a good portion of the Earth's atmosphere out of harm's way below the observer. Finally it turns out that this atmospheric refraction effect varies a little according to the wavelength of light, so that the star image is not only displaced upwards by the atmosphere, but spread out vertically with the blue end of the spectrum uppermost. This is also the part of the spectrum which is most scatted by interaction with the molecules of the air. The best cure for this particular problem is to restrict the sensitivity of the photographic apparatus to a narrow range of the spectrum, judged to give the best record of the star field, and to arrange to determine effects such as differential atmospheric refraction from one part of the photographic plate to another, by a scheme of comparison exposures. Even though at high angle in the sky the refraction effect is small it differs from one site to another according to the local air density and humidity, the rules for this being pretty accurately known. Since the eclipse operation aimed at the highest possible precision nothing which might impair this could be ignored.

Next, we come to the lens which is to image the star field. Ideally, in some respects, the best lens is no lens at all, simply a pinhole. Every point in the image sends a ray to the pinhole which passes through it and impinges on a photographic plate at the rear of the camera. If we hold the developed plate up with the eye at the same distance as the pinhole from the plate we shall see a perfect directional image of the sky. However it will be extremely faint since only the tiny area of the pinhole received any light from the object. So we replace the pinhole with a lens of a certain area, to act as a greater catchment of incident light. But this now has a certain focal length. The lens receives a parallel bundle of rays from each point in the object and brings them to a focus at a certain distance from the lens. As we swing round in different directions from the centre of the plate the focal points trace out a sphere, so that, ideally the image of the sky, which may be treated as a sphere since we are only concerned with directions to sky features, is formed on a spherical surface. When we replace this with a plane photographic plate we find that the object topography has been transformed systematically to a new topography on the plate. This has all been known for a very long time, and such transformations have been worked out and long applied as standard features of astronomical photography. When a simple lens with spherical surfaces produces an image the parallel rays which pass through the outer zones of the lens come together at an axial point nearer the lens surface than parallel rays coming through the more central parts of the lens. Strictly speaking, there is no true focus, but a small recognizable region within which a point source will be imaged as a small circular blur. Worse still, every lens acts not only as a deviator of light rays, but also as a weak prism, deviating rays of different wavelengths by different amounts and producing the aforementioned 'focal' region at a slightly different position. Long ago the situation was improved by the discovery that a two-component lens, one convex, one concave, of different glasses could be designed so as to minimize these effects. Such, so-called 'achromatic' lenses were not truly indifferent to color, but could be designed so as to treat two chosen wavelengths equally, resulting in a compromise optimum arrangement.

With the introduction of miniature cameras before World War II, and the burgeoning scientific and commercial demands since then, lens designers have found it possible to devise ever more complex, multi-component objectives, with vastly improved compromises—compromises are inevitable—in the final result. One additional problem introduced by this new complexity is that clean glass surfaces act as weak mirrors, and a multi-component lens may have several unintended ghostly reflective images inside it, unless something can be done about them. The remedy came along with the growth of what is now the vacuum deposition industry which can apply quarter-wavelength-thick transparent coats to glass to inhibit reflection of a pre-determined wavelength of light.

At last we come to the photographic plate, where, as we shall see, important technical consultations with the extremely obliging research chemists at Kodak, took place. Plates would be of glass of certified flatness to avoid distortion, coated with an emulsion restricted to a particular spectral sensitivity range chosen for best results at the eclipse.

## Sage Advice from Princeton

In the spring and summer of 1972, there were intensive discussions between Texas and Princeton, many carried on by correspondence, or over the telephone, with several taking place in Princeton, with Bryce attending or sometimes taking the chair. Perhaps the most important document is a detailed letter dated March 9, from Dicke to Harlan Smith giving a review of the photographic and instrumental problems which might have to be surmounted. This was still at the time when Texas was contemplating three separate expeditions, with specially manufactured instrumentation, for which specifications would have to be written. On the subject of instrument design, Dicke wrote "Concerning coelostat vs. movable telescope, I would guess that the tube length and mechanical difficulties may be controlling here. With new mirror materials the old objections to mirrors are reduced. One unconventional system worthy of consideration is a telescope looking at the pole and rotatable about its optic axis with a single mirror fastened to the tube (and rotating with it) giving the declination." This would have been an interesting development, but with hindsight, one cannot fail to remark that the time available for new instrument development was very short. If Texas should indeed produce such an instrument, or rather three identical ones, Princeton expressed a wish to be an inheritor of one of them after it was no longer needed for the eclipse.

In writing about these problems, Dicke, of course, peripherally mentioned most of those expounded in rather more detail above, since they are all well-known, often by bitter experience, to scientists engaged in this field. He then went on, especially in relation to Al Mikesell, who had been trained as an astrometrist, to discuss such topics as possible distortion of the emulsion coating of the plates and ways of controlling it, as well as ways of checking the performance of the telescope, of whatever kind might be chosen at the time of the eclipse. One of the important things would be trying to control the temperature of the telescope and lens since, as we already know, in general an increase of temperature would cause an increase in the focal length of the instrument, leading to a degradation of the star images and a change of plate-scale, one of the most sinister enemies of any Einstein-shift observation. The subject of calibration went back and forth several times, starting with a proposal to include on a primary eclipse plate an exposure of a sky area just to the west of the eclipse and another just to the east, both at the same altitude above the horizon as the eclipse plate. The subject of calibration exposures is apt to make the flesh of the expedition-planner creep. The length of the total phase of the eclipse is all too short—though in the present case at six minutes and eighteen seconds it was near the maximum possible—and it seems positively reckless to waste some of it on anything but the eclipse itself. One also has a slightly illogical funny feeling about daytime exposures of plates, but one should reject this, because it isn't daytime and the sky is pretty dark all over during totality. Lastly, one has visions of a star-studded sky, and thinks that these extra exposures on the same plate will saturate it with lots of images—how can we tell which is which?—with some comparison stars actually sitting on top of wanted images from the eclipse plate. Actually these fears are unwarranted: the standard sky surveys have pinpointed the relevant stars, and there are really not too many of them. A quick glance at a typical astrographic plate will often give the impression that there is nothing on it at all, until one picks up the surprisingly sparse images. The calibration proposal began to get even more complicated with triple exposures proposed for each plate, and, what was more portentous, a steady increase in the size of photographic plate required. Here, a certain amount of prudence would have been more appropriate since the plates, to avoid risks in returning them to civilization, would have to be processed on site, and the larger the plates—they ended up as 12 inches square, each weighing 3½ lbs—would have been increasingly difficult to process perfectly under Saharan field conditions, to say nothing of the fact that, under human stress, the more complicated a procedure the greater the risk of fatal error. Lastly, Dicke had a proposal for an artificial field of perfect round images, which might be impressed on an eclipse or calibration plate by contact printing. This was to take a glass plate and strew on it—or place systematically—a series of very small plastic spheres, available commercially, before putting it in an aluminum coating plant. Afterwards the surface could be gently brushed to remove the spheres, leaving a distribution of tiny round holes, which would produce images about 2 arc seconds in diameter of perfect pseudo-stars. There was also a discussion of the plate-measuring machines which had recently come into use, but this topic can be deferred until we come to the actual eclipse.

All this intense activity was dealt a blow by the decision after the receipt by the National Science Foundation of the original Texas-Princeton proposal, which they had clearly been found much too expensive, at about $400,000. The project would be supported, to the tune of about $65,000, but would not include funds for new instrumentation. It was this decision which caused Bryce to write to the IAU representative, saying that only one site would be chosen, and Mikesell to inform other interested parties. The proposal for polar telescopes thus fell by the way, with the response from Princeton that they would not now take an active part in the expedition, though they remained available as a source of friendly counsel if needed. There was also a certain stringency of funds, including those for payment of staff, as the result of which, the French astronomer, Maurice Marin, who had been a colleague in Texas on a different instrument project returned to his own country.

## The Search for Equipment

If no serious instrument novelty could be entertained, a search began to find suitable equipment on loan from among a galaxy of friendly institutions. There were many astrometric lenses in existence, whose properties were well-known from particulars given in publications where they had been used.

However, some were not particularly suitable, others were installed in equipment still in use, which their owners did not want upset, or were about to be used in new projects. The search ranged over American observatories such as Lick and Harvard, to optical manufacturers with long-standing friendly relations with the astronomical community, such as Perkin-Elmer in the USA and Taylor, Taylor Hobson in Britain, to South Africa where enquiries were made of the Cape Observatory and possibly the Leiden observatory station in South Africa. Success came through another French optical expert, Jean Texereau, like several others a friend of Gérard de Vaucouleurs, who had been called in to improve the figure of the 82-inch secondary optics. (Evans & Mulholland, 1986). He revealed that, stored in his workshop, was an astrometric lens made by a specialist French optical firm, (REOSC), for André Danjon, former Director of the Paris Observatory, but never used by him. Texereau put Harlan Smith in touch with the current Director of the Paris Observatory, Professor R. Michard with a request to borrow the lens for the eclipse.

This was a four-component, blue-corrected, flat-field, lens of 20 centimeters diameter and 2.1 meter focal length, weighing 30 kg, installed in a tube as an astrographic unit. In requesting the loan, Texas promised full insurance coverage for the period of twelve months over which Texas custody would be needed, and asked permission to do some drilling on the telescope tube. This would have applied had the Princeton proposal for a polar telescope been continued. The response was that, if this were borrowed, it must be returned in exactly its original condition. So only the lens itself, in its mounting, was borrowed, and the insurance went down to $1600 for a capital sum of $18,000. In transit it was always carried by Bryce De Witt as personal baggage. In Austin, after the component elements of the lens had been anti-reflection coated for a wavelength of 4400 Å, a wavelength at the extreme blue end of the visible spectrum, the lens was stopped down to a diameter of 16.5 centimetres to reduce general sky brightness on the plates. In addition, a Schott GG-395 flat glass filter, specially ordered from Germany, also anti-reflection coated, was added, and the lens adjusted with extreme care over a period of three months, mainly by Dick Mitchell, who wrote a long instruction manual for its care and installation. As Burton Jones remarks in the first published paper on the eclipse results, tests showed that after hand-carriage to Chinguetti the adjustments had not deteriorated. It left Austin on May 12 1973 as part of the personal baggage of Bryce and Cécile. Packed in a metal container with handles, with a second compartment containing the photographic plates for the eclipse, the whole thing weighed 80 kilograms (176 pounds).

The lens was of a type known as Petzval after its originator. The image of a star-field formed by a lens, lies in a plane some distance behind it, but, with small departures of image formation from ideal circularity, increasing with distance from the axis. (Aberrations). If one turns the lens round so that the rays go the opposite way through it, they still come to a focus in some plane behind the lens. So if one uses first a lens (in this case a two-component doublet), and then sends the rays on to an almost exactly similar one reversed, the second lens does not undo the convergence of the first, but does have the possibility of diminishing the aberrations of image-quality produced by the first lens. This was the design principle behind the construction of the borrowed lens. It did involve what turned out to be a real head-ache, that the two doublets were in a metal mount, subject not only to expansion and contraction with change of ambient temperature, but even to inequalities of temperature at different parts of the metallic fitting, so that the whole thing had to be carefully heat-soaked to achieve uniformity. This would prove particularly trying in the Sahara where enormous daily changes of temperature were common, even between different areas quite close together. None of this was fully realized beforehand, and the planners congratulated themselves on having acquired the use of a high-quality piece of essential equipment.

With a suitable lens in hand, Texas began looking for a mounting for it. This was an item discovered within the Texas system, though at the time located just outside Austin, at a site chosen for visitors and perhaps some initial telescope training for entering students.

This is a story in itself. Julien Péridier was a well-known French engineer, who made a large fortune in 'the public transportation industry'—including the Paris Metro. He was also an enthusiastic amateur astronomer, who observed a solar eclipse in Spain in 1905, and reported many observations of variable stars to the appropriate sections of the French and British astronomical associations. In 1933 he set up a considerable private observatory near Le Houga, a small town in the south-west of France, almost within sight of the Pyrenees foothills. The result was a considerable output of serious astronomy done by a number of visiting students, including work in the field of photoelectric stellar photometry. Among the instruments was a 12-inch diameter Newtonian reflector made by the Victorian English optician, George Calver. (Unsigned 1928). He made a large number of mirrors, including a 37-inch diameter mirror for his friend Dr A. A. Common,(Dyson,1904), which, with its mounting, was subsequently sold to Edward Crossley, who after using it himself, presented it to the Lick Observatory, where it had an honorable and distinguished history. Calver died in 1927 aged 92. Péridier bought his own 12-inch from Calver and installed it at Le Houga. (De Vaucouleurs, 1968) The senior author's first exposure to a telescope of any size was to a duplicate of this one, installed in a little show-place observatory in his home town of Cardiff, Wales.(Figure 42).

Among the astronomers to work at Le Houga was Gérard de Vaucouleurs, including a period during the German occupation of the northern half of France, who became a close friend of Péridier. De Vaucouleurs had been one of the first recruits to the Austin astronomy faculty, even just before the arrival of Harlan Smith. His patron died in 1967 at the age of 85,

Figure 42. The telescope restored to
its original form after the eclipse.

Figure 43: The telescope as modified
for the eclipse expedition.

bequeathing many of the resources of the observatory to de Vaucouleurs, including the 12-inch reflector. This telescope tube now became the support for the eclipse lens. It was turned from a reflector to a refractor with the installation of a plate-holder mounting, and a new drive, by the in-house construction experts, under Johnny Floyd, as described in more detail below. (Figure 43). A comic, and unverified thought, has occurred to the senior author. When Calver designed this tube and mounting, with the usual drive and counterweights, he must have used screws following the British Association standard system of threads. If Péridier did anything with it, he must have installed metric threads. When Johnny Floyd's experts took it in hand they must have used standard American threads. The stand now sits unused at what we call the Bee Caves site, but some machine archaeology might be amusing.

This old warhorse was now to undergo a kind of sex-change operation, albeit only temporary, from reflector to refractor, ready for its two excursions into observational astronomy in its new guise. This involved the provision of a mount for the lens to hold it rigidly in position (the total front end load had now reached 75 pounds) and a plate-holder assembly, also heavy, to secure foot-square glass plates a quarter of an inch thick. The front end incorporated not only a sleeve to hold the lens in position, but also a dew cap. The rear end was to incorporate a sector just in front of the plate center. These are time-honored devices for reducing the intensity of parts of astrometric plates which might be too intense for comfort. In this case the intent was to reduce the intensity of the image of the eclipsed Sun so as to make its surroundings more comfortably represented on the plates. The traditional so-called sector is a disk mounted on an axle, and put into rapid rotation, often by a little air-flow turbine. The disk has sectors cut out of it like a statistician's pie-chart, symmetrically displaced so as to obviate vibration. If, for example, each open sector were 18° wide, then when the disk was rotated in the incident beam, the area behind it would be exposed for ten per cent of the exposure time, and thus subject to a calculated intensity reduction.

The telescope drive was to be replaced by an accurate electric friction drive. To control expansion and thus to maintain a pre-determined focal length, the lens assembly and some other parts of the equipment were to be thermally insulated, and provided with heating elements and thermisters to define exactly their thermal state. A mount for a guide telescope was also installed.

## A House for the Telescope

It was already decided that the plates should be processed on site, and that the equipment be left sealed after the eclipse for several months to allow comparison plates of the eclipse sky-region to be taken at night. All this demanded that housing be provided for the telescope, capable of resisting bad weather—though clearly sustained rainstorms were not a likely possibility—possible human interference—though the known high standards of ethical conduct in this Islamic state meant that nothing worse than boyish larking might be a danger—and affording facilities inside for controlled temperature processing of the pretty large and somewhat unwieldy photographic plates. And, of course, the housing had to be taken to the desert and installed there.

Matters began with a brainstorming session at the Architecture Department, and included proposals for a geodesic dome, and surprisingly, a house made of compressed cardboard. Before writing off the participants as having taken leave of their wits, one can consider the undoubted merits of this idea. The structure had to be sufficiently strong to last for several months, but there would almost certainly be no rain of any duration. It could be sealed against incursion of dust, would be intrinsically rather well insulated, and would probably not weigh much, an important consideration if it was to be shipped to the site by air. In the outcome a more conventional structure was devised by engineer Allen Brune. Mikesell gave a good description of it in a letter to Philippe Crane at Princeton, on June 13 1973. He wrote :—"Picture of the telescope housing. View from the NE : the door is on the S side. Base is 16 x 16 feet, height to flat roof 12 feet." (Actually the upper part was a truncated square pyramid). "The eclipse hatch is half the flat roof. Material 1/4 inch plywood, cemented to 3-inch reprocessed Styrofoam. with 2x 4 studs on 4-foot centers. Specific viewing hatches were provided in each slant roof, but holes could be opened most anywhere for viewing the sky. Required about 14 packages for components, when shipped, and weighed about 3500 lbs, cost, outside labor and materials about $ 2000. We assembled it as pictured "(in a parking lot now occupied by the Petroleum Engineering Chemistry Department building), (Figures 44,45), "in less than three days. On site it would take three people over a week to complete, dust and light tight. Temperature control is with an evaporative air cooler, (giving 3000 cubic feet per minute maximum flow to give filtered air and a positive interior pressure), and a 1.4 KVA compressor-type air-conditioner both together requiring less than the 250V 90 Watt output of our small on-site generator."

"Inside there is a double-door arrangement for light and dust-locking, and a 4 x 12 foot dark room alcove with temperature controlled to 75 ° F. Ambient external temperature range—in the shade (in the Sahara)—is 70° at sunrise, to 105° at mid-afternoon. Internal temperature is thus well controlled. The paint outside is the standard $TiO2$ white which astronomers favor. Inside is aluminium foil over white. There is no floor. The ground is insulated with foil and fiber-glass and remnants of shipping crates". The date of this letter, and the nature of the report makes it clear that this was an account of the performance of the hut after its installation on site in Chinguetti, not yet connected up to the NCAR base installation to be put there, and reported by radio back to Austin. The reader may complain that the account here presented has a tendency to jerk capriciously from one time-frame to another, rather than sequentially, a form of exposition more or less necessitated by the fact that a whole series of separate preparative actions were going on simultaneously.

It must be realized that it was no mean feat to descend on a corner of the Sahara and to install a structure capable of functioning like a precision chemistry laboratory in an environment menaced by dust and extremes of temperature, even though every component was numbered and a careful instruction guide provided. The hut had removable panels as required by the observational situation, and was to be erected with edges running east and west, with an observing aperture to the east. This was because on June 30, a date just after the solstice, the Sun's declination would be 23 ° 13 ',less than three degrees different from the site latitude of 20 ° 27 ', so that at eclipse time the motion of the Sun would be almost exactly vertically upwards, as it headed for a later close near-zenith passage at noon. The hatch allowed for access to a patch of sky at the correct altitude of about one hour in Right Ascension. It always comes as something of a surprise to less than close students of geography, how far to the west the bulge of Africa extends. Well within this, at longitude 12°22' W, Chinguetti would receive maximum eclipse at UT 10h 48m 50s, with the Sun to the east of the meridian at the altitude of 61°. The hut was emblazoned with an inscription in Arabic which announced that the American astronomers came in peace, a choice of wording so similar to a quotation from the Koran that it ruffled official feathers among NCAR personnel, fearful of unintended implications of the inscription among the locals. It was intended, after the eclipse, to lock up the hut to seal it against incursions by sand, until a follow-up expedition could come in to take night-time plates. A guard would be hired to keep a benevolent watch over it. After it was no longer needed, and the inside equipment removed, the hut would be

Figure 44: The hut assembled on a
campus parking lot.

Figure 45: Bryce awarded an accolade in the completed
hut, by Charles Thompson and Al Mikesell.

donated to the local community. In due course it was reduced to its components and shipped by ocean freight via Dakar to Nouakchott and Chinguetti.

## NCAR Takes Charge

It was enormously helpful to the Texas expedition not to be left to provide all its own logistical support, but to participate in the larger effort provided by NCAR. As we have seen, this organization was issuing an extremely efficient series of instructional communications, with which all the participants were required to comply. No 2 signaled the choice of sites, that at Chinguetti very close to the position on a sketch suggested by Bryce, not far from the gîte, and one in Kenya near Lake Rudolph. No 3 dealt with travel plans, which included transport by chartered jet to Senegal. Bryce had already nominated his team, including himself and his wife, who would travel independently to Chinguetti in mid-May, as also would Burton Jones from his base in Britain. Alassane Sy, (usually called Gorel by his family), as the only Mauritanian, would not use that service either, but come from his base in France. From Austin would come Richard Matzner, an assistant professor in the relativity school at the Physics Department in Austin, Richard Mitchell of the astronomy staff, and Charles (Chuck) Cobb an experienced night observer from McDonald Observatory. These last would be flown to Dakar in Senegal, housed temporarily in a local hotel, (the SU-NU-GAL), and then flown by local aircraft, first to Nouakchott, and then to the airstrip at Chinguetti.

Communication No 4 was a short course in environmental protection for instruments and personnel, and an elementary account of the history and customs of Mauritania. The population of the capital, Nouakchott was given as 55,000.

No 5, issued on January 19, gave details of the shipping requirements for transport by air, with such things as package sizes, and shipping marks for the various destinations. All equipment, which had to be at a shipping agent's in New York between April 2nd and 12th, would then be flown to Dakar, and off-loaded for truck transport if intended for West Africa, with the rest going on to Kenya aboard the same charter aircraft, for similar handling there. The packing requirements included the cautionary statement that material to be returned to the USA would have to be re-packed in the same crates as those used on the outgoing trip.

The items destined for Mauritania bore shipping codes M-30 to M-41, but four of these were reserved for their own use by NCAR, and some were consigned to Atar, so that, in the end Texas shared the Chinguetti site only with Kitt Peak, Harvard, Florida, Hawaii and New Mexico. There were 23 items shipping-marked for Kenya. Later it was emphasised that the packing lists had to be in French, no problem certainly for Texas. Communication No 7 named the physician for the Chinguetti site, a Dr Warren Gillette from Boulder, Colorado, who advised that participants should start their regime of antimalarial pills. Several communications dealt with such questions as insurance—designated the responsibility of each expedition. More importantly Communication No 10 gave a very long and detailed list of the equipment to be provided on site by NCAR, such as a hoist and post-hole diggers, to avoid unnecessary duplication by individual expeditions. The NCAR installation would include two HF and VHF generators which, when the camp was fully established, would supply power to the individual sites, to be laid out according to an accompanying diagram. (Figure 46). At eclipse time, one generator would be running, with instant readiness to switch to the other in the event of failure. It was at this point that Mikesell registered the requirement of the Texas group for 10 to 20 gallons of water per day to operate the expedition's evaporative cooler. There would be no dry-ice available on site, but Texas needed six one-foot square concrete blocks, four inches thick.

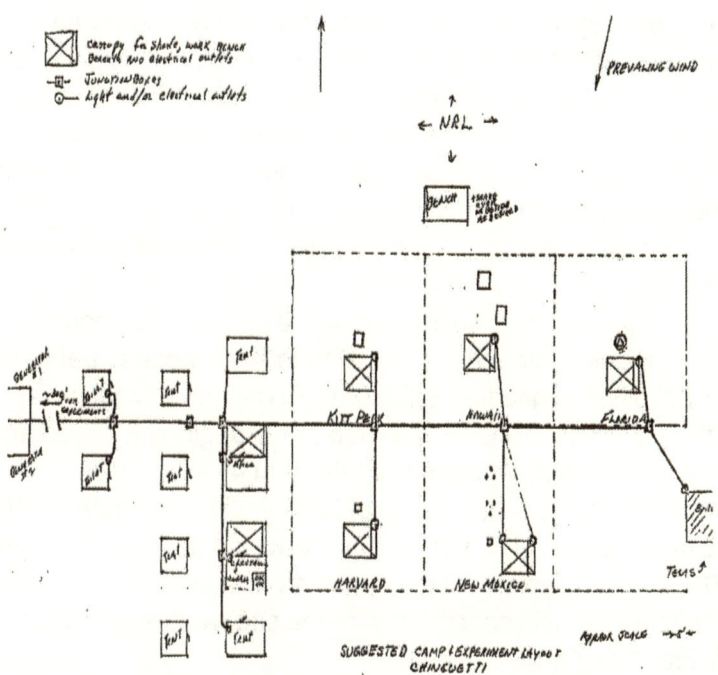

Figure 46: Layout of the arrangements for
the six scientific expeditions headed for
Chinguetti, as proposed by NCAR.

- 181 -

Various other requirements were pointed up in correspondence from Mikesell to Medrud. Then came Communication No 13, an almost avuncular set of instructions, covering the finances of participants. They would receive cab fare, or car mileage at 12 cents per mile to get them to their local airport, with an open ticket provided from there to JFK and return. From the moment that travel began, they would get $10 per day ($ 2.50 per quarter) as far as the observing site. Hotel accommodation was arranged and paid for by NCAR at Hotel SU-NU-GAL or at N'GOR in Nairobi. On site, food and lodging would be provided by NCAR, plus $ 2.00 per day per head for incidentals. Personal freight was limited to 50 lbs.

For the charter flight to begin on June 6, it was noted that the onward segment from Dakar to Chinguetti had very little space available and scientific freight must be kept as low as possible.

Communication No 16 B gave the eclipse circumstances for Chinguetti, specified as, altitude 480 meters, Latitude 20°27'.0 N, Longitude. 12° 21.'9 W, (Bryce actually had values differing by fractions of a minute. The beginning of totality (second contact) would be at 10h 45m 41.4 s, maximum at 10h 48 m 50.9 s,and the end of totality (third contact) at 10h 52 m 00 .5 s, all UT. Partial eclipse would begin at 9h 28m 19.1 s. A detailed plot of the lunar limb was provided. (Figure 47). The scientific coordinator for the site was designated as Dr Frank Q. Orrall of Hawaii.

## Address to the Troops

Following a good naval tradition, on April 16, Bryce drafted a before-action message to the troops who would be with him at Chinguetti. He did not know all of them particularly well, and he wanted to establish good relations, especially in view of the fact that they were going to a demanding physical situation, which would try the patience of some. Not only that, they were engaged in an enterprise which required meticulous attention to detail, a slight loss of which might defeat the goals of the whole project. Had it been necessary, he might have had to seek to control an errant member of his group with only the persuasion of his personality, and no real disciplinary authority. In the event, nothing like that happened since every member was a willing volunteer, intensely devoted to the success of the project. Nevertheless, this was a very wise document.

His first paragraph deserves quotation at length:-

"This is my first communication to you in my official capacity as director of on-site operations for the eclipse expedition. Most of my communications in the future will, I hope, be verbal, in the field with lots of feedback from you. It is my fervent desire that my official role be as inconspicuous as possible, and indeed I shall depend on you all in laying out our work program. I think that we shall have little difficulty in deciding collectively our order of work and priorities, but occasionally I may have to make hard or even unpopular decisions. We have a responsibility to the University of Texas, to the National Science Foundation, and to the many other persons, institutions and agencies that have made this costly expedition possible. Therefore I must ask you to conduct yourselves, on site, with quasi-military discipline. I wish I could say that discipline can be forgotten during off-duty hours. But, even then, we shall have a responsibility to our hosts, the people of Chinguetti, in whose care we shall be leaving our equipment after our departure, until the follow-up team arrives in November."

His next paragraph dealt with four topics. First, neatness, systematic organization of operations and the duty to leave everything afterwards in shipshape condition. It may not be clear to those who have never worked under these conditions, but, for example, let it be said that he who puts down at random a screwdriver in the Sahara, hundreds of miles from all possibility of replacement, and diverts his attention to other things can bid adieu to his screwdriver and everything he wanted to do with it.

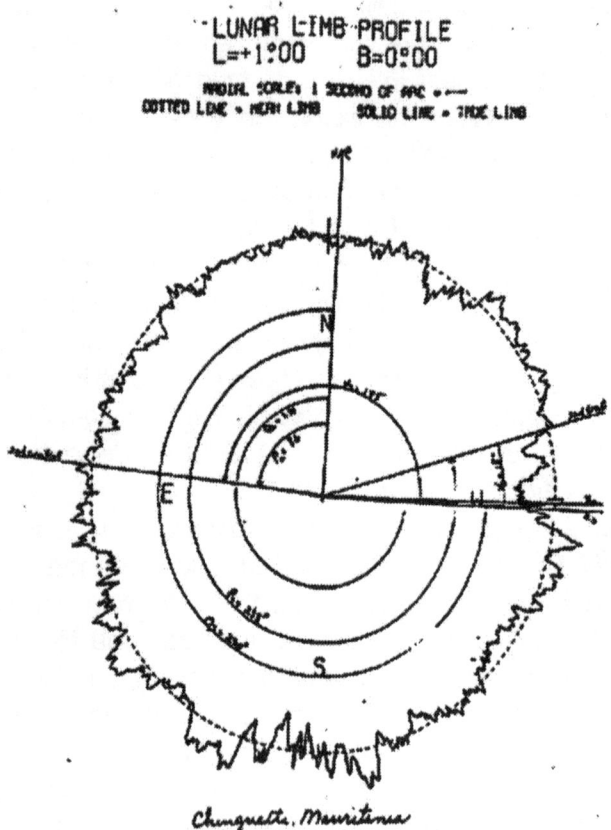

Figure 47: A plot of the lunar limb profile
expected at eclipse time supplied by NCAR.
The radial scale of deviations from the circular
form is enormously exaggerated. The plot is
reversed as compared with Figure 3, so that
the exceptionally large feature producing the
observed Baily's bead appears at the right.

Secondly, Bryce insisted on the precise execution of tasks, unhurried even in stressful conditions, and with no inspirational variations unchecked by himself. Those of us who have operated in the field have all, perhaps, recollections of the disastrous consequences of last-minute bright ideas, involving deviations from an agreed routine.

He wanted everyone to keep a notebook or running log, with a dated record of work done, including data records, calculations, observations, and most of all, immediate identification numbering of exposed photographic plates.

Lastly, "until we have ascertained the precise nature of our impact, and that of the other teams, on the water supply of Chinguetti, water is to be used sparingly."

Bryce concluded by airing two possible work periods in a day's schedule, one in the earlier part of the night, the other later. Both were tentative, and were arranged in relation to meal-times at the gîte and the need to have cool daylight for some jobs. He hoped to organize a meal break in the middle of the work period, whenever that might occur. He promised to declare a holiday if people began to get exhausted, even though they had a tremendous amount of work before them.

He cautioned that there would not be room for more than two, or at most three, people, to sleep at the same time in the shelter. They would have to sleep on the floor, but the place would be air-conditioned and relatively quiet. Perhaps there would have to be a schedule.

He looked forward very much to joining the rest of the team in Chinguetti, whither, of course, Burton Jones, Cécile and himself, would be the advance party.

Cécile had prepared a short document, entitled "An Extracurricular Activity of the Advanced Eclipse Team of the University of Texas at Chinguetti (Mauritania)" from which we quote.

On the one hand she hoped to establish good relations with the local school system, which, following so soon after the end of French rule, was still set in the traditional mold of that of metropolitan France, with which she was, of course familiar, and to participate in some instructional activities.

She also quotes the statement that Chinguetti was the ancient capital of Mauritania, the site of a madrasa—a center of Islamic instruction—and a library. "Like all Islamic cities of learning it has traditionally offered generous assistance 'to those who travel in the quest of knowledge'".

"Although the Chinguetti library has been mentioned to me by several Mauritanians as the best collection of Arabic manuscripts in the southwestern parts of the Islamic world, very little seems to be known about it. I have talked to several specialists, (including Derek Price, David Pingree and Owen Gingerich), who have given me strong encouragement, but very little indication of what to expect. I have looked into the possibility of identifying old eclipses and their usefulness in determining the acceleration of the Earth's spin" (Newton, 1969; Fotheringham,1920-21).

The Governor of the Seventh Region of Mauritania, capital Atar, Baham Ould Mohamed Laghdaf had expressed a special interest in the library. She had also become acquainted with Professor Alassane Sy, a professor of physics at the University of Orléans, who had expressed his willingness to help the Texas team. Though not a specialist in Arabic manuscripts, Cécile hoped, with the assistance of Alassane Sy and perhaps the local teachers, to make a preliminary reconnaissance of the Chinguetti treasures and to spend a happy time with its keepers, especially if there should turn out to be some items with implications of astronomical history.

## The Advance Party Leaves

Finally, Bryce and Cécile were off on May 12 for New York, by Braniff Airlines, then Air Afrique to Dakar and Nouakchott, to return by Air Afrique to Paris on July 2—or at least that is what they intended—but their schedule suffered

Figure 48: The suitcase carrying the lens came from the 1971 occultation expeditions—but with new labels.

Figure 49: The solar-powered water pump resembled an English Parish Church.

some changes. Helped by Dick Mitchell, and Al and Marjorie Mikesell, they presented themselves at the Austin check-in counter with nine pieces of baggage totalling a little over 500 lbs. This included one item which was to travel with them in the cabin, all the way to Nouakchott, the metal box containg the 8-inch lens, and some 30 of the best photographic plates, weighing in total some 150 lbs. This got a separate seat, where it would be tied down by the belt. Bryce's idea was to put the heaviest of the baggage on their two-items-per-passenger allowance, and pay excess on the rest. He matched Cécile's personal suitcase with a large case which had been used on the 1971 Jupiter expedition to carry one team's computer to Australia. Following that effort there were six large metal cases as souvenirs in the Department, and this was one of them. (Figure 48)

They were met at the ticket counter by Mr M.H. Benjamin, Supervisor of Ground Operations at Austin, with whom they had discussed eclipse plans the week before. He hoisted the Australian case on the scales, took one look at the weight, (115 lbs), and waved all the rest through with no further comment. The baggage was quite picturesque, being covered with labels announcing its provenance from the University of Texas, its relevance to the forthcoming eclipse in the Sahara, and warnings against exposure to X-rays. There was trouble at security, because Al had brought a collection of forgotten tools, and security forbade Cécile to take them into the cabin, so they were handed over to be checked. Bruce and Benjamin carried the lens box up the stairs—this was at the old airport in one of its earlier manifestations—and up to the front seat in the first-class section, where it and the DeWitts were lodged, and duly accorded the treatment then customary, by courtesy of Braniff, even though, they and their charge, had only economy tickets. A Braniff agent met them at New York, and helped them get to the Air Afrique terminal, where they were pleasantly treated, but had to pay for their excess baggage. Here they had to wait while a shipping agent found a customs agent, who fixed up the documentation to allow the French lens to come back into the USA without paying duty. They also met Claude Morel of NCAR, like them headed for Chinguetti, whom Bryce had last seen in his hotel in Niger.

The flight to Dakar was uneventful, but they would have to spend some time there, since the next onward flight to Nouakchott would not leave for 36 hours. They were duly lodged in one of the hotels satellite to the big N'Gor chain and enjoyed the stopover on the beach, the swimming pools and the restaurants. Before they were picked up by the Air Afrique agent next day, they were joined by Burton Jones, who had been in one of the other satellite inns. At lunch, he mentioned that van Altena, at Yale, recommended using Perma Wash for developing plates, which obviated the necessity of keeping them in the wash water so long, with subsequent risk of emulsion creepage. Bryce cabled Al asking him to investigate this material.

After dealing with the bond for the lens, they proceeded to the airport, where Burton Jones and Bryce slipped under the barrier and accompanied the big lens box to the plane. At the ladder he was asked whether they were diplomats, and when he answered in the affirmative, the box was carried on board and stowed against the front bulkhead. They then returned to the terminal to pass through customs, where a man came to Cécile and said that the pilot of the Caravelle objected to the location of the box. A suitable location was duly found. The pilot got interested in the eclipse, and she sat up on the flight deck drinking champagne, while Bryce and Burton sat with the other passengers.

At Nouakchott, customs inspection was complete chaos, with all the luggage piled in a heap to be identified by the owners. The big box was to one side, and Bryce and Richard McCullough, NSF's Nouakchott agent, who had come to meet them, quietly picked it up and took it to a waiting car.

The rest of the baggage was loaded into a Land Rover,

and they proceeded to the hotel Marhaba, where Burton, Bryce and Cécile checked in. McCullough kindly sent round his rented car for their use. Cécile called Dr. Ba, whose full title was Dr. Ba Bocar Alpha, who wanted to know what they were doing in that hotel, and invited them to supper. They were embarrassed at leaving Burton in the lurch, but McCullough and Morel took him under their wing. Dr. Ba had the reputation of being the best doctor in Mauritania and the one with the most palatial house. At that house the DeWitts met all sorts of influential and highly placed individuals, including Dr.Touré, whom they had last seen in Washington, who, like Dr. Ba, was an uncle of Alassane Sy : an aide to the President of Mauritania: and a former Ambassador to Tunisia. Dr. Ba gave them a goatskin *guerba*, (water carrier) and would try to get Cécile an interview with Moukhtar Ould Hamidou, an elderly specialist in Mauritanian manuscripts. McCullough said that getting an invitation to the Ba's was the envy of everyone in Mauritania.

Thursday May 15 was a busy one, with running to the bank, contacting Brahim Danabja to see if he could locate their old and trusted driver, Mahmoud, visiting the pharmacy and American Embassy, ordering a complete native costume for Bryce in the market, and calling on the office of the Poste et Télécommunications to arrange for Richard Matzner's ham radio license. They had lunch at the Ba's and brought Burton along for supper. Cécile was to see the manuscript expert the next day. There was also a long discussion concerning the arrival of Alassane Sy in June. It was hoped to have NCAR's assistance in getting him on the Chinguetti charter on June 8.

It seemed that some difficulty had arisen concerning land travel to Chinguetti, there now being an officially-backed consortium of Land Rover owners, who had the sole rights to this service for eclipse personnel. So they could not use the trusted Mahmoud, except by the subterfuge that they were only going to Chinguetti to look at manuscripts.

There was a real duststorm on the next day, Wednesday May 16, which so filled Dr. Ba's courtyard that the servants gave up trying to clear it. Cécile went off with Dr. Ba on some medical calls, and then to see Mr. Moukhtar Ould Hamidou, while Bryce and Burton went to pick up Matzner's ham radio license. While there, they had a chat with Mr. A. Duffau, head of the radio communications section, himself a radio amateur. During the eclipse he would emit a continuous signal at 14.050 MHz, so that any amateurs who wished could report if the eclipse shadow affected reception.

Cécile had spent a fascinating morning with Mr. Ould Hamidou at his office in the Palais de la Culture, (built by the Chinese Government), which also contained the National Library. Mr. Ould Hamidou was a scholar in the strictest sense of the term, and while he searched for references to old eclipses and comets, he hovered over Cécile, making sure she dotted every *i* and crossed every *t* in the material she was reading. He had given her the names of several families in Chinguetti, who owned the manuscripts in the so-called Chinguetti library. Most of them were in private hands and jealously guarded, but Hamidou would like Cécile, if the family members agreed, to make a complete photographic copy of the oldest Koran in existence, never before photographed. Bryce wrote to Mikesell asking him to send a better lens for his camera. It may be noted that this claim was disputed by one of the Arabic manuscript experts consulted by the present authors, and, indeed, it did not surface again during the eclipse period.

McCullough arrived to ask about their plans for traveling to Chinguetti. As Mahmoud didn't seem to be in Nouakchott, they decided to travel to Chinguetti, all in one day, by Land Rover. McCullough had brought the consignment of pure water, 30 liters which Bryce had asked for. They invited him to lunch, along with themselves, knowing how relaxed the Ba's were, and especially that Mrs. Ba wanted to meet him. They were hailed on the street by Mr. Sidi El Mockhtar

N'Diaye, a former President of the National Assembly, who had received the grant of independence from the hands of Charles de Gaulle on Mauritanian Independence Day. Mr Sidi had often been at the Ba's, sometimes asleep in a corner while the others talked. He was accompanied by the new Governor of Atar, Mr. Yarba Ould Ely Bourba, (replacing the previous Governor Mr. Laghdaf), whom Mr. Sidi had brought along to introduce to the DeWitts. The Governor invited them all to lunch the next day at his residence in Atar. He would not be there himself, but would notify his staff to look after the party.

At the Ba's they found that Dr. Ba had left for Boutilimit, a desert town south-east of Nouakchott, to attend funeral rites for the father of the President of the Republic. Because of the furious duststorm he could not go by air, and the overland journey was a long round-about one. Islamic tradition required the body to be buried before sundown. At the Ba's McCullough had an excellent chance to talk with Dr. Touré, Mrs. Ba, and Dr.Ba's brother, an enthusiastic fisherman, as also was McCullough. The latter was invited to accompany him to the wharf as often as he liked, for a little sea-fishing.

About 4.00 p.m. they returned to the hotel to repack their luggage, to be called on by representatives of TRANSECLIPSE to enquire their desires concerning a second Land Rover—and who was going to pay for it. In the negotiations, Cécile secured a larger vehicle with more seats. At 6.00 p.m. Dr. Ba's brother arrived on his way to the wharf for some fishing, so the DeWitts and Burton Jones squeezed into his Volkswagen and drove the three or four miles to the sea. Because of underground seepage it had not been possible to build the town any closer. At the wharf, there was heavy surf, with several ships anchored several hundred yards out. The lighter crews were quitting for the night, and their vessels were being hauled up by cranes on to the wharf. Within ten minutes Dr. Ba's brother had caught several fish, 10 to 12 inches long.

On the way back he pointed out the de-salination plant which supplied Nouakchott with fresh water. They all supped at the Ba's and returned to their hotel about 10.30 p.m.

They were awakened at 5.30 a.m. the next morning by the arrival of McCullough and Morel, who were all ready to go, but it took so long to get the bags packed in the Land Rovers that they only got away an hour later. In the end, Bryce left the plates in the big box, and held the lens on his lap all wrapped in its protective packing.

Cécile, Bryce, Burton and Claude rode in the vehicle with the seats arranged like a station wagon, with McCullough in the other one. Bryce saw at once that their driver was incompetent, young, and cocky, with minimal driving experience. The engine timing was bad, and they could not go faster than 80 kph, and on top of this, the other Land Rover had an engine failure, bringing everything to a stop. The driver of the second Land Rover diagnosed distributor trouble, and the personnel fooled around for an hour with no results. McCullough decided to go back to Nouakchott with the first Land Rover to demand a new one from the TRANSECLIPSE pool, whose contract specified that vehicles should be in good repair, and inspected before departure.

While they waited, two cars came past, one with a man from Geneva who wanted to sell them some tracts, the other with the director of the SOMIMA mine and his wife, both English. They asked for a message to be sent to Atar from their destination at Akjoujt, explaining that they would be late for their lunch appointment at the Governor's residence. Unfortunately this message was never delivered. At 9.30 a.m. McCullough returned, not with a new Land Rover, but a new distributor. After half an hour the driver still had not got it installed, so McCullough told the others to go ahead, their vehicle being the one containing the delicate equipment, and he would follow somehow.

They reached Akjoujt at noon, in an atmosphere so dusty that visibility was less than five miles. They had lunch at the SOMIMA restaurant, where they were overtaken by McCullough. There, most unusually, they were addressed by young man in a boubou, who accused them, in French, of coming to Mauritania to get money to build houses in Britain, supplemented by an insult in Hassaniya Arabic, at which all the nearby kids and the driver laughed.

Because their Land Rover made such poor speed, it was 5.00 p.m by the time they got to Atar. Instead of going directly to the Governor's house, the others wanted to go first to the gîte, to see if there might be rooms for them there. There were not, the gîte being in progress of refurbishment for the crowds expected the next month. A nearby building yielded five beds in a dismal place with no running water but plenty of insects.

At this point, a man appeared who said that he had been waiting for them with a meal since noon. This was the Commissaire de Police who had received the message from the Governor the day before. They apologized meekly for their delay caused by their Land Rover troubles. At this, the Commissaire said he would have the vehicles inspected to make sure that there would be no more trouble. At the Governor's residence, they were ushered into a large salon, and served the usual three glasses of tea by the servants. Except for a sideboard at one end there was no furniture, but a large rug and five or six mattresses on the floor. An old-fashioned electric fan stirred the air. While the tea was being served, Cécile slipped out to call on the Governor's wife. In accordance with Moslem custom, she would not come out to greet the men, but she called Bryce back later with Cécile to say hello, and to verify that he really was her husband. More apologies were offered, together with, for her husband, a picture of the Earth seen from the Moon, which they all signed. The Governor's wife said that they were welcome to stay overnight, provided that the DeWitts slept upstairs, and the rest in the salon.

Before supper the Commissaire reappeared. They presented him with a Chinguetti button after explaining the plans for the observations. They asked if he knew Mahmoud Ould Amar Cheine, their chauffeur from the previous year, to which he replied that Mahmoud had been in his office half an hour before and that he would go and fetch him. This he did while they were at supper, telling Mahmoud only that some friends of his were at the Governor's residence. When Mahmoud arrived he was astonished to see them, and the DeWitts rushed to embrace him. He immediately invited them to his house after supper for tea. In due course they all piled into his Peugeot and drove to his place near the edge of town.

When they arrived in the unlit alley, the family were waiting to see who he was bringing, and on seeing the DeWitts all started shouting, hugging and calling Bryce the grandpère. Bryce carried the littlest girl, (six years years old), into the kerosene-lamp-lit courtyard where they sat or lay while Mahmoud prepared the tea. The glasses were laid out in a semi-circle on a brass tray, the brazier was burning, and Mahmoud began the familiar steps, breaking off bits from the sugar cone with a special hammer, stuffing mint into the pot, ending by pouring the tea back and forth, from a height of two feet, from glass to teapot, until each glass had its froth on top.

There were hundreds of things to talk about after a year's interval, and although both Claude and Richard could join in the conversation, while Burton had to sit on the sidelines, the whole scene was so rare that he did not feel deprived. They eventually asked if the girls were as good dancers as they had been the year before, in response to which they then proceeded to put on a far better show. The eldest girl fetched a metal washbasin and beat a syncopated rhythm on it, while singing lines of song. These were repeated by the two littlest girls, who stood facing each other, moving their feet, hips, and hands, in a rehearsed choreography, while the rest clapped their hands in time with the complicated beat. Even the littlest boy, who had obviously been circumcised since they had last seen him, beat out the rhythm on a can of fruit brought by Cécile as a present. The visitors were able to respond with a variety of small gifts, including a wallet and a purse, some Chinguetti buttons, copies of photographs taken the year before, and some small pieces of jewelry, saved for years for just such an occasion. McCullough, who had been in Mauritania for six months, said that this was the first time that he had been in a Mauritanian household, except for formal occasions at the Ba's with only men present. He drove them back to the Governor's residence and promised to see them off the next day.

By now it was Friday May 18, and the DeWitts had resolved to travel to Chinguetti with Mahmoud, having learned that the young kid who was driving their Land Rover had never been to Chinguetti before. Bryce could not bear the thought of the plates being left to bake in the Sahara Sun or the lens being subjected to violent bouncing. This made it necessary to lie to the others. Medrud's orders had been that they were to be brought to Chinguetti in two Land Rovers, and they could not oppose his wishes on this. The government contract with the TRANSECLIPSE organization made it illegal for anyone else to carry eclipse personnel. So they had to say that Mahmoud would take them for the price of gasoline alone. This deception had to be backed up by Mahmoud, and while Cécile was telling McCullough of Mahmoud's generous offer, Bryce was explaining to Mahmoud that they would pay the standard price, out of their own pocket, with no necessity of giving them a receipt. They would, all the same, hope to be reimbursed from the National Geographic grant, which did not require receipts, and would be represented as a disbursement in furtherance of the research on the ancient manuscripts. Bryce's ex-post-factor justification was the report of Burton's Land Rover ride, with his head more than once banging on ceiling bars, circumstances such that the lens would never have got through unharmed.

Although the Land Rovers had set out first, the Peugeot soon passed them, making frequent stops to verify that they were still coming. The lens suffered not the slightest bump, even on the climb up the Amogjar pass. Only within the last 25 kilometers to Chinguetti did the sand start to become really bad, and Mahmoud had repeatedly to swing off and back on to the road while keeping his momentum through the sand. Here Bryce hung on to the lens on his lap for dear life. Eventually even Mahmoud lost face by getting stuck. The trucks and Land Rovers had torn up the road so much recently that he no longer knew exactly where to jump off the road and head across country. He almost missed one critical point, and finally missed another right by the airstrip 7 km from town. Even by deflating his tires and with the DeWitts pushing, he could not get moving again, and the others were needed to extricate him.

## Chinguetti Revisited

They found the gîte thoroughly modernized, with electric lights, air conditioners in every room, a bar and a dining room with canned music. The earth-packed courtyard had been replaced by a gravel-covered area with a fountain in the middle, and a manhole cover over a pump which lifted water to a cistern on the roof. The illumination of the cloisters at night prevented the stars from being seen.

Medrud and his assistants, Dick Bobka and Bob Bowie, were just coming in from the eclipse site as they arrived at noon. They gave him the same explanation for Mahmoud and the Peugeot. When the Land Rovers arrived he gave them all a résumé of the living and dining arrangements, and they then all went in to lunch, bringing Mahmoud with them. For an expedition into the desert, this was certainly high living. The

Société Hotelière de Ravitaillement Marine had done a bang-up job of bringing modern comforts to Chinguetti. Not only had the gîte been repaired and repainted, but a dilapidated building alongside had been renovated, providing more air-conditioned rooms. A big generator ran day and night, breaking the peaceful silence that this part of Chinguetti once possessed.

The water-pump for the gîte was powered by solar energy. There were said to be only five others like it in the world, one of them located at the solar energy center in Niamey, which the DeWitts had visited the previous year. The device was invented by a professor in Dakar and was operated by an enthusiastic young Frenchman. To a later eye the installation bore a remarkable resemblance to a village church in rural England. (Figure 49).

The kitchen of the gîte had several large refrigerators, so one of the first things Bryce did was to put the photographic plates in one of them, wrapped in aluminium foil provided by Dick Bobka, to prevent their being wet by any of the liquid contents of the refrigerator. The plate boxes showed evidence of their suffering severe jolting on the ride from Atar, but there was no sound of broken glass, only of the plates or their spacers slipping back and forth when the boxes were tilted in different positions. The master réseau seemed undamaged.

At about 3.00 in the afternoon Medrud took them out to the site to show them around. He and his assistants had done a superb job of getting things ready and organizing work crews. The whole camp was laid out with the tools in order in various boxes, tents, or small buildings. The two big generators were in operation, two flag poles had been set up, and the men were busy setting up a large transmission antenna. Bryce's first reaction to all this richesse was that Texas had brought too much, a sentiment strengthened by the sight of the boxes all lined up, dwarfing those of the other expeditions.(Figure 50). He then thought with alarm of Al

Mikesell and Richard Matzner, who would be all alone in Chinguetti in November. In fact, of course, they would not be.

## Preparing the Site

After this, the three wandered out to the Texas site to decide exactly where to put the building. The ground was flat, but rocky, at one side of the soccer field. The basic allocation of space had already been decided by NCAR, close to the spot chosen by Bryce, and laid out by them. On this sort of ground, utterly innocent of vegetation, the schoolboys had calisthenics, (Figure 51), or played soccer in the mornings. The work began with having all the building materials brought over to the site, including the telescope base, which was hauled behind a Land Rover on a timber sled. The local laborers, supplied by NCAR, tried to improve the site by removing some of the rocks.

Mahmoud came to the gîte after supper and invited them to have tea with some relatives of his. They walked a short distance through the night, and came to a courtyard, where a number of robed figures were seated on the ground in the moonlight. Two of them were government inspectors come to inspect the local school. Couscous was being served and they were invited to join in, which they did, and, though they were very full from supper, ate a little. The standard way to eat hot couscous was to take a handful and keep shaking it in the hand, to form a ball to prevent it being so long in contact with the skin that it burned, and then to transfer it to the mouth. Tea was served not long after, and one of the inspectors tried to teach them some words in the Hassaniyah Arabic dialect. The pleasant evening ended with Mahmoud walking them back to the gîte, where he had been invited to stay, using the other bed in Burton Jones' room. He had said that he would like to see what it would be like, sleeping in an air-conditioned

Figure 50: Bryce was abashed to see that Texas had more baggage than all the other expeditions combined.

Figure 51: The schoolboys exercised or played soccer on the bare ground.

room, but later, Burton, who had gone out with Claude Morel to make stellar observations with the theodolite, reported that when he got back Mahmoud had disappeared. Next day Bryce, overtired, got up late, to learn that Cécile had got up to say goodbye to Mahmoud, who had left on his own before the departure of the pick-up type Land Rover which was taking McCullough back to Nouakchott. Mahmoud was carrying down the same Percepteur who had ridden with them to Atar the previous year, thus regaining the face he had lost on the upward journey.

The crew began unpacking the crates, sheltered by a canopy provided by Medrud, with Cécile keeping track of everything and maintaining some sort of order. The attempt to clear the site of rocks having failed, they had the workers bring in several Land Rover loads of sand from a nearby dune, to level the area. (Figure 52).They intended to lay a plastic sheet over it, and then woven mats purchased in the village, such as were used for traditional ground-cover for ordinary domestic life. By the end of the day they had the base-plates for the hut bolted togther, with a small trench dug for them and the foundation level. The direction was determined by compass, allowing 11° W variation. Bryce was discouraged by the weather, visibility being much worse than a year ago, and there were even some cumulus clouds.

Next day was Sunday May 20, and working hours had settled to be either from 6.00 a.m. to noon, or from 6.00 p.m. to 7.30 p.m. They had been consuming enormous quantities of bottled water, plus an occasional salt tablet. Bryce had even drunk some of the water from the gîte. NCAR would supply them, free of charge, with all the pure water needed for photography. This day they had put in position the concrete pads needed for the telescope base, and leveled them, and the base, in its box, had been put on top of them by a local work crew. By the end of the day all the sides of the hut were up, with the middle ones bolted together and screwed in position on top. They felt a sense of accomplishment, in spite

Figure 52: The designated site for the hut.

Figure 53: The Préfet with two Mauritanian soldiers.

of having paid a formal call at the house of the Préfet, followed by lunch there at 12.30. (Figure 53). This turned out to be an enormous affair of course after course, followed by the usual three glasses of tea. The discussion was in French, with occasional interruptions for translation for the benefit of Medrud and Burton Jones. By the time lunch broke up at about 3.00 p.m., they were so sleepy they could hardly stand, intensified by stiffness brought on by unaccustomed postures, stretched out on mattresses on the floor. Nobody turned up for work until 4.00 pm., except for Bobka and Bowie, who had been working on the radio. They made first contact with Kenya that day, and would try to reach Boulder on the morrow. Keeping the laborers happy was a problem, a job delegated to Cécile. Though there were only two assigned to the Texas group, it was a problem finding work that the group were willing to let them do, because Bryce and Burton preferred to do the hand-tool work themselves. So there wasn't much beyond lifting and positioning panels, left for them to do. Cécile had them laying out panels, fetching things from the other end of camp, clearing rocks, and piling rock and sand round the skirt of the building.

At the end of the day, a young boy arrived, and said that they were expected for supper in the evening at the home of some of Mahmoud's relatives, but Cécile went to their house to explain that pressure of work and fatigue made it impossible to accept that night. All the same, she had to stay for tea. While there, she picked up various items of information of differing levels of credibility. The Préfet had decreed that any villager approaching the gîte or the eclipse site would be fined Fcs 6000 AFR, which explained why there had been no curious onlookers: even the little messenger boy had, at first, been turned back from approaching them. The Préfet had already assigned a 24-hour guard to their own building at a cost of Fcs 38000 AFR per month, to be paid by NCAR, and by Texas after June 30. Less credible was the conclusion by the townspeople that Burton Jones was the DeWitt's son: and that they were going to be handing out uniforms to everybody—this the result of Cécile's giving the Préfet two eclipse pencils, two Chinguetti buttons and one eclipse T-shirt at their formal meeting.

Supper at the gîte was uneventful. The gérant, (manager), ate at a separate table at the same time as the visitors, and Cécile engaged him in conversation because he seemed a bit lonely. He was a nice enough guy, who had had an interesting life, but kept asking, when Bryce was not around, when she was going to invite him into her bed.

Next day, they got the building squared up to the same dimensions as it had in the Austin parking-lot, and began to put the roof on. They had very little use for the laborers at this stage. At noon, instead of going directly to lunch, they all went to have a look at the solar-powered water pump guided by the young Frenchman in charge of it. Its large roof was covered with pipes fastened to black-painted sheets of metal. Water circulated through them and exchanged heat with a two-piston engine propelled by a mixture of oil and butane, which could exist either in a gaseous or liquid phase. There was a condenser to convert these phases. The water was pumped to the gîte, and later, would go to other parts of the village. There were some camel watering-troughs nearby. They were told that the water-level in the wells was lower than in the previous year, some 35 feet below ground level.

Cécile began to feel unwell and was ordered by Bryce to stay in bed. That left only Burton and Bryce on the site, except for the guard, who sat under the canopy telling his rosary beads. That day they got all the corner pieces of the roof in place, as well as the top plates and two of the rectangular roof panels. Burton was very pleased with their progress, but Bryce felt they were moving extremely slowly, because all the observational work still lay ahead. By the next day, they had expected to have the hut completely insulated against dust, caulked, and weather-stripped. It had not been too hot to work, being reminiscent of the summer season in the Central Valley of California, back when Bryce was pitching watermelons.

Bryce felt very fit and got to work next day (May 22) at 6.00 a.m. after a good night's sleep. The standard schedule for the NCAR group was from 6.00 a.m. to noon, with a break for breakfast at 7.30 a.m. This meant driving back to the gîte, but the gérant couldn't get his kitchen staff to work any earlier. The afternoon shift was from 3 or 4 p.m. to 6.30, but Bryce liked to stay on for up to an hour, to use the last bit of daylight, and then would walk back to the gîte. That morning, they put up all of the roof except for the top panels and inserted all the metal strips between the roof and the walls. They stuffed the empty spaces under the strips with glass fiber for insulation.

Medrud's people working the radio were still unable to contact Boulder, but would keep trying. So the visitors had begun to feel out of things. Mail delivery was scheduled only once a week, and the NCAR group had received only two letters in all the time they had been in Chinguetti. The weather continued good with a light breeze and good visibility, though not up to last year's standard. The Sun, at rising and setting, was still very pale because of dust in the atmosphere.

Cécile, who had arranged a laundry service for everybody, was still not quite well and did not come to work until the afternoon. At the hut they finished putting up the outside metal stripping and caulked the outside seams on the lower half of the building, so that one of the laborers could pile dirt round the outside. Burton and Bryce stayed on while Medrud's people were trying to contact Boulder on the radio. There was so much noise and interference from a station in the Spanish Sahara, that it seemed hopeless, but at 8.00 p.m. both Kenya and Boulder came through loud and clear, and Bryce was able to transmit an encouraging message for relay to Mikesell, reporting progress, and asking that the main party in June bring some nails and some exposed black photographic film. Everybody was in a good mood and talked late after supper.

Mr. Anezin, the gérant of the gîte, had been very nice and stopped making passes at Cécile.

Next day, they got to work early and made good progress on the hut. Cécile still felt a little unwell, and retired early. Although the inside temperature was seldom lower than 77° F, the outside temperature got up to 106°F or 115° F, depending on which thermometer was consulted.

Cécile's trouble seemed to be in adapting to the changes from the air-conditioned interior of the gîte.

After some more detailed work on the structure, they opened the box containing the telescope base, unbolted it from the box-floor and used the hoist, provided by NCAR, to put it in position on the concrete slabs. All this by the end of the morning.

At lunch they found that Mr. Rafael, overseer for all the sites that the Société Hotelière de Ravitaillement Maritine was remodeling at Atar, Chinguetti, and Ouadane, (this last the ultimate oasis beyond Chinguetti), had arrived. A nice, but tough, guy, able to overcome all the difficulties that arose in an undertaking of this kind in those surroundings.

In the morning, three camel teams of two or three animals each, had passed not far from the hut, carrying loads of what seemed to be wood. The combination of the camels, the date palms and the giant dunes (Figure 54), in the background, was so picturesque that Bryce took his camera along in the afternoon, but, of course, no camels showed up then, and he had to content himself with a few shots of the hut and his colleagues having tea with the guards. Cécile had brought a supply of tea, six glasses, a cone of hard sugar and a daily supply of fresh mint, which she had turned over to the guards, asking them to make tea two or three times a day, for tea instead of coffee breaks. Everybody sat around the charcoal fire at tea time, which took half an hour for the statutory three glasses, a bit longer than the average coffee break. Apart from this, they worked from 4.00 to 6.30 p.m., and got the telescope base mounted on its metal pads, aligned as well as possible using the magnetic compass. They finished up the day with more caulking, and getting the inner door, and antechamber mostly installed. They worked late because the NCAR people were also staying late to operate the radio. They made contact with Boulder again and exchanged lots of messages, as well as questions and answers. One was a message for Cécile from Mikesell which they could not decipher. Bryce remarks, en passant, in his diary that he had noticed that the man who was guardian of the gîte the previous year, who had shown them Prantner's unsigned traveler's checks, now worked for the Préfet.

Cécile had a very bad night of intestinal upsets, but bravely went back to work in the morning. She filled a very useful role as guardian of the 'stock room', and finding things for the others. They were in real need of her services this day, and by the end she was somewhat better, but very tired. Bryce had planned a whole series of tasks for the day, but got none of them finished, only some progress on the inner door and antechamber, and the air cooler. Mr. Ould Daddah, the Ministre de l'Equipment was expected in Chinguetti this day, but never showed up.

The next day, Friday May 25, was much more successful. With Dick Bobka's help they got the air cooler operating. It was mounted on four concrete blocks with a fifty-gallon drum sitting on top of it. Bobka soldered the copper tubing to the drum, and Saidu, supervisor of all the laborers, brought 100 liters of well water to put in the drum. The float and water flow still needed some adjustments, but the equipment put out a fine blast of cool moist air, though Bryce thought it would be better with the spare motor pulley to reduce the air flow somewhat. The antechamber was now so airtight that it was hard to close the doors. The day was windy, with much fine sand in the air and generally reduced visibility. It turned all the white strips of caulking pink before they dried. It was the prevalence of this fine sand that made Bryce determined to get the hut airtight before they dared start assembling the

Figure 54: The giant dunes seen in the distance.

Figure 55: A stroll in the nearest palm grove.

fine components of the telescope. Cécile was much improved in health, though not entirely free of unpleasant symptoms. Morning and afternoon tea had become a regular part of their lives. Just after siesta, a young boy came to invite the DeWitts to tea, explaining that his was a very fine family of *marabouts*, (local upperclass citizens), and that an invitation to tea also implied an invitation to supper. Cécile explained that the invitation would have to wait for a few days because of pressure of work.

Bryce remarked in his diary that they had been on the site for a week, had lost count of the days, and could only get straight by counting back the days of his entries. This day there was some contact with the outside world: the arrival of two letters sent via the American Embassy in Nouakchott. The others who had used the official expedition address had received nothing.

During the morning Mr. Ould Daddah and his entourage arrived in the town, but left immediately for Ouadane before the DeWitts could see them. They returned in the evening and went for supper at the Préfet's house. Cécile left word that they would be delighted if Mr. Ould Daddah would inspect the site before leaving town again. In anticipation of his visit, Cécile had the Arabic sign mounted on the wall next to the door. This was 12 feet long and two feet high, with an inscription drafted by one of her students, which said, roughly:-

"The University of Texas is grateful to the people of Chinguetti for their renowned hospitality to those who come in search of the secrets of the Universe."

Later that day, Mr. Ould Daddah with his retinue, including the Préfet, the Percepteur and Jean Abdallahi, came from the Préfet's house to visit the site and expressed their admiration, especially for the Arabic sign. After a good siesta, Bryce and Burton worked on some of the electrical installations, including those needed to connect to the NCAR generators. Cécile had the plastic cover down on the floor, and they swept and and vacuumed the whole building before laying the mats bought from the villagers, on top of the plastic.

They were still tinkering with the air cooler, which needed various adjustments, since the air flow was so good that unless the roof hatch was left slightly opened it popped out from the excess pressure. The evaporative cooler worked marvelously, because the ambient relative humidity was only a few percent and evaporation went on at an express rate. However, it seems that they had yet to realize the problems presented by a water-guzzler in a region where that element was rare and precious. Maybe that began when another hundred liters of water had to be added to the system this day. Bryce wanted to defer the installation of the air-conditioning system in the darkroom, until the telescope installation had been completed.

The Préfet came with the Chairman of the village council to have supper with the whole gang, except for the two working on the radio. After a very agreeable evening, Bryce and Cécile took a stroll to the nearest palm grove. (Figure 55) He promised himself a holiday once the telescope was in operation, which he proposed devoting to an exploration of the oasis, which was actually spread in an arc several miles long.

Beginning the assembly of the telescope next day (Sunday May 27), Bryce and Burton were frustrated by the lack of three unique bolts involved in the installation of the polar shaft. In their high and low search, they brought in all the other cases of equipment, which had to be vacuumed on being opened. They eventually found two of the bolts in a box of miscellaneous oddments, which had been standing in their tent for days.

Meanwhile, Cécile had gone off with one of the laborers, who lived in a nomad-style tent, with his wife and family nearby. They took with them one of the mats which she had bought to receive instruction on cutting them, necessitated by the peculiar shape of the telescope building, thereafter to be dignified by the title "observatory". The problem was that the mats could only be cut in one direction, which left a lot of loose ends. Normally these were stiff and brittle. The laborer took the mat, buried the cut end in the sand and poured water over it. After ten minutes, the cut ends were pliable, and could be woven back into the mat using two special hand-forged needles that the labourer supplied. She also went to the old part of the village accompanied by one of the Land Rover drivers, to buy leather sandals to replace her rubber ones, which were being cut to pieces on the rocks, some desert clothing, and some abominable-looking lots of seed pods, which, when soaked in water, produced, according to the driver, a drink good for picking up from fatigue after a day's work.

In the afternoon, Bryce and Burton encountered more of the sort of trouble they had had with the special bolts, which turned out to be too short for their task. They were rescued by Dick Bobka, who had a complete set of tap-wrenches. Bryce noted, as we have above, that the telescope was a confusing hybrid of English, American and metric threads. Bobka also got the air-conditioner into a stable operating mode. Without his aid they would have been in real trouble on this frustrating day.

Next day, slowly but surely, they made progress on the telescope, getting the tangent arm, and its box, counterweights, tube cradle and tube, guide telescope and plate-end of the tube, all in place. The declination axis and its screw gave a lot of trouble in ensuring that the telescope would move freely without binding. Dick Bobka again came to the rescue when they could not find missing parts. They had trouble with fine sand blowing in through the roof opened wide to accommodate the hoist, so that, afterwards, everything had to be vacuumed again. They could not take time off for tea, so Cécile brought it to them, in, of course, the customary three shifts. She was spending a lot of time with the laborers and the guard, who were teaching her some Arabic. She could rattle off some of the standard incantations :—"Allah ilaha il Allah / Mahomedon rassou lalah", with the greatest aplomb. Bryce hoped to be able to release her to get acquainted with the teachers after one more day of work on the equipment. They intended to install the lens next day.

It is appropriate here to note some of the problems encountered by the use of the telescope as an eclipse instrument. In its original manifestion as a reflecting telescope of Victorian vintage for use in European latitudes, it would have been mounted on a cast-iron pillar, with a polar axis, at an inclination no doubt adjustable for a relatively small range of latitudes in the European range, a cross-axis with counterweights on one side to balance the tube on the other, which turned around this axis to access star fields in different declinations. The drive, round the polar axis to compensate the rotation of the Earth, and thus maintain fixed the desired star pointing, would have been a carefully machined clockwork, driven by a falling weight mechanism with a mechanical governor. A clamp was provided so that the instrument could be set at a particular chosen declination, while a clutch mechanism could release the telescope from the drive mechanism round the polar axis, allowing it to be pointed to the east or west to the desired field, at which point the tube assembly could be re-engaged with the continuously moving drive mechanism.

At a European latitude of, say, some 50 degrees, the polar axis would have been tilted up at this angle, and when the telescope was pointed at some field on the meridian, the cross, (declination), axis would have been horizontal. So a swing from low to higher on the declination axis would have corresponded to a movement in the direction south to north. And a swing from left to right about the polar axis would have corresponded to a change in the east-to-west direction.

For the eclipse situation these movements would have been almost interchanged. The polar axis would have been only a little reared from the horizontal, to accommodate the latitude of 19 degrees of Chinguetti. So, when the telescope was pointed towards the well-risen Sun in the East, moving almost perpendicular to the horizon, a swing from lower to higher would correspond to an east-to-west movement, and a left-to-right swing would correspond to a north-to-south movement. To accommodate this situation, the engineers had constructed a new support system to go on top of a welded-steel tripod support. The gravity drive was replaced by an electric drive, provided by an Accutrac motor operating on a tangent screw.

The tube, originally in the form known as a Newtonian reflector, had been built with a cell at the lower end containing the telescope mirror, and with a diagonal reflector at its upper end to bring the sky image to an eyepiece near the top end of the tube, at one side, at a convenient height for a standing observer. In its manifestation as a refractor for eclipse use, the mirror cell had been replaced by a photographic plateholder assembly. The spider holding the diagonal reflector on the central axis of the tube at its top end had been removed, and so had the eyepiece assembly. At the top end, an extension to the tube carrying the heavy lens borrowed from France produced, when properly adjusted, a focused image of the sky on the photographic plate. The original telescope had only a very small finder telescope mounted parallel to the main tube and directed thus to the same sky area. Now something more powerful was needed to pick up a suitable guide star not far from the eclipsed Sun, and so to verify that the main instrument had been correctly pointed. It would seem that the engineers had done their best with all this modification, but would have preferred to have had more time to carry it out. For a time the Questar telescope was mounted as a finder. (Figure 56). Plans to install the lens the next day, (Tuesday May 29), were thwarted by the arrival of gale-force winds blowing fine sand in the air, which necessitated battening everything down and covering all the telescope parts with plastic sheet. Bryce decided to postpone work on the telescope and to concentrate on finishing off the interior details of the building including the photographic darkroom. They started work on a mounting box and duct system for the air-conditioner but they made little progress in the bad weather conditions.

Mr. Ould Die, Secretary-General for Tourism spent the night in the gîte, and had said that he would come to the site at 8.00 a.m., but he never showed up, possibly on account of the storm. When they went back to the gîte for lunch, they were so covered with fine sand, which even got into their mouths and onto their teeth, that they had to shower first. As Bryce remarked, the gîte was by then so well equipped that they had no need no worry about the storm raging outside. After supper, they worked late on the air-conditioning system, running a separate heavy-duty line for it to the NCAR generator. Cécile meanwhile had been in touch with the local teachers. By then the radio was in touch with Boulder and a message was sent congratulating the DeWitt son on his graduation from high school.

Although the wind died in the night, by next morning it was back in full force, demanding attention to final caulking and insulation of the hut. In the storm the guards sat patiently with their *haoulis* (turbans) wrapped around their heads so that only their eyes showed. It was impossible to make tea. The visitors returned to the gîte and slept all afternoon, finally getting to work until past midnight, in calmer air so full of dust that no stars could be seen. It was reported that rain had finally come to the southern region of Mauritania, near the Senegal River.

The gîte was temporarily short of water, the well being deliberately pumped dry so that a dynamite explosion could increase the flow. May 31 was devoted to cleaning up after the storm, with Cécile organizing a tool-rack, after which they unpacked the refrigerator and put it in the dark-room area. They felt that at last they had the place reasonably well organized and proof against outside storms.

Figure 56: The Questar, still mounted as a finder on the upper end of the telescope after restoration to its original form.

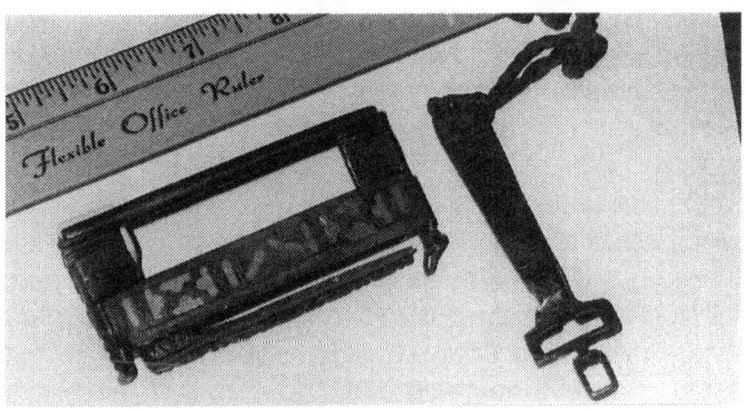

Figure 57: Cécile had bought a locally made padlock, with a separate key, whose operation defied explanation.

At 5.45 p.m., one of the Land Rover drivers, a very high-class nomad type military veteran, came to take them to a *méchouie* that he and his friends were preparing for them. The NCAR trio were also invited, but declined, saying that they had, on Prantner's instructions, to stay in the radio-shack to talk to Kenya.

The ride to the site of the méchouie revealed to Bryce, for the first time, the true dimensions of Chinguetti. They traveled two or three miles until they reached the very last, picturesque, outpost, of the village. Bryce put on his Mauritanian clothing causing quite a stir. The guest list included the Percepteur with his wife and two little girls, and the school teacher who had looked through the Questar on top of the gîte the previous year, and asked so many questions about the satellites of the planets. They were first given a tour of the area with many ruined buildings covered by the encroaching dunes.

At the palm grove, the méchouie, a sheep wrapped in cloth, buried in the sand, with a fire burning directly above it, was being prepared. Nearby a rifle range with four longish stones at a distance of 200 yards had been prepared for their entertainment, and everybody was allowed one or two, extremely scarce, cartridges, fired from an old French army weapon, (strictly contraband). A great carpet was spread next to the well under the palms, next to the pool of well water, with masses of mint growing on the banks of the pool. With the aid of only fingers and knives, the sheep was consumed, followed by a large bowl of the herbal infusion that Cécile had tried to make a few days before. Everybody drank out of the same bowl, and then there was tea. After that, everybody piled into the Land Rover and rode back to the gîte. Only Claude and Burton were in European dress, Cécile also having put on her local clothing. She was carrying a bunch of mint picked at the site of the feast, and offered it to the guards, who, at once started a tea-party, during which black clouds came up, there was even a distant flash of lightning, and a few drops of rain fell.

Next day, up at noon, they devoted their energies to a general clean-up, unpacking crates, moving everything into the hut, using the empty crates and their lids to provide necessary shelving space. When Bryce produced some of the balloons he had brought, and blew them up, the guard and the two little girls started chasing them across the sand unavailingly in the wind. Cécile took ointment to one woman whose baby had some kind of eye infection, and found herself compelled to treat the whole family, even though most of them were not sick.

In the late afternoon, the weather turned bad with cumulo-nimbus clouds, strong winds and clouds of dust. The visibility was terrible and even though the winds died down at dusk, the only star visible was Arcturus. Just after sitting down to supper Bryce was called away by the Land Rover driver, who said that the Minister of Public Health, the Governor of Atar and other dignitaries had arrived and were anxious to see the eclipse equipment. Bryce found himself in the center of a score of blue-robed figures, whom he showed round the building by flashlight, the power being off. They seemed pleased and impressed by the size of the telescope and expressed their thanks for the Arabic sign. In due course, they set off in two military Land Rovers intending to reach Atar that night. When Medrud, Claude and Bryce returned to the gîte a light rain was falling, and lightning, accompanied by thunder could be seen in all directions.

The first discovery next day was that the building was not completely dustproof, and that some rain had got in, though it had not been strong enough to wet the ground appreciably. This called for a lot of work improving the various joints and seals, until 8.00 a.m., when the school teachers arrived with 42 children in tow. Cécile let them inside in small groups, and then gave them an astronomy lecture, while Burt and Bryce showed off the various gadgets.

They had hoped to get the lens in position during the morning, but were held up by all sorts of small unforeseen problems. They did manage to get the aluminum back-plate of the telescope etched with phosphoric acid and primed with zinc chromate, in preparation for a coat of black paint the next day.

At lunch time, strong winds got up, yellow clouds boiled everywhere, and a light rain began to fall. They did hear that there had been a cloudburst in Ouadane, which filled a wadi and carried away a camel.

The weather had been so cool during the previous few days that Bryce and Burton had been skipping their siesta and working through the afternoons. Cécile had occupied her time with errands in town, seeing the Préfet about the library, and about laundry, and buying soft shoes for the team to wear inside the building, so as not to tear up the woven material of the mats, made from the sort of coarse grass to be seen growing among the dunes.

Bryce and Burton began work with the lens, only to discover that Dick Mitchell had put the thermistor 180° from its correct location on the inside surface of the lens. This they rectified, and increased the size of the cardboard aperture to 6 7/8 inches following Mikesell's instructions. After finally mounting the lens they went to install the dewcap, only to find that the special screws needed had been misplaced and were only found after a long delay. They finished the afternoon by balancing the telescope.

At 6.00 p.m., Mr. Jacquet, the manager of the solar pump, came to conduct the NCAR and Texas gangs to the house of his assistant, designated to succeed him, for tea. They enjoyed lying around by the ground-level windows, bargaining for objects that their hosts were anxious to sell them. Burt and Jacquet bought a considerable quantity of the arrow and axe heads with which the Sahara abounds, and Cécile bought an old Arab padlock operating on an unusual principle, really only explicable with the object available. (Figures 57, 58, 59). A child with an eye affliction was brought for Cécile's attentions.

Figure 58: Quite simple!: Just put the loop on the
key over the rod, and force it down.

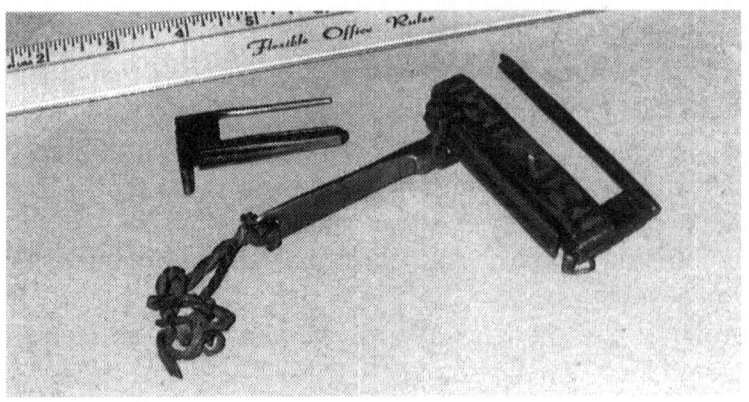

Figure 59: The padlock easily separates into two parts.

On the way back to the gîte they were stopped by the young son of Mahmoud's cousin, who complained that they had not kept their promise to go for tea and supper, made on their arrival at Chinguetti. Bryce stopped off long enough to change into his Mauritanian dress, and they then went to a meal of the usual staples :—couscous, camel meat, eaten with the hands from the same large bowl, preceded by a *zrig* cocktail, (diluted and sweetened goat's milk), drunk from the same large cup. Both the boy and his sister had been with the group visiting the observatory that morning. They were very bright, describing local customs and trying to teach them some Arabic words. The rain had cleared the sky, and the stars shone brightly as they sat in the courtyard by the light from a single kerosene lamp.

Next day, Sunday June 3, they completed the installation of the telescope, including the tangent-screw drive. All went perfectly and Bryce declared a holiday for the next day, not only as a celebration, but as a beginning of the change to night-shift work.

Cécile again performed as a physician. It seemed that the antibiotic eye ointment did some good, as judged by the improvement of a child of one of the guards. She also had a long talk with the Préfet at his house, mainly centered round astronomy, in which she gave him a short course.

He said he would take them to see the Arabic manuscripts the next day.

That day, Monday June 4, was the promised holiday, starting with a lecture by Cécile to the class who had visited them the day before, and backed up Bryce and Burton over a 2-hour question and answer period. (Figure 60). Photographs were taken during recess, at which Bryce met the camel boy of the previous year, to whom he gave copies of pictures of him taken in the dunes.

They then walked over to the old part of town, where they met a man with arrowheads to sell, of which Burton bought a large quantity. On the way to lunch, Cécile had to stop to administer some more ointment to the daughter of the guard with the infected eye.

After lunch, they went to look for the Préfet who was to take them to the Imam, who was, in turn, to take them to the library. The Préfet was tied up with some people from the Mauritanian broadcasting system, who had dropped by the gîte in hopes of getting an interview with the visitors, but sent an aide who took them to the mosque. The Imam was a short, elderly, bald, bearded, wiry man, with a very kind face, dressed in coarse dark cloth of the plainest kind. (Figure 61). His mosque, which they were not allowed to enter, had a minaret no more than 20 feet high, but simple and neat, and a courtyard where travelers might spend the night. (Figure 62).They could view this from across the alleyway which served as a street. The library was a couple of dingy storerooms off a courtyard not far from the mosque, piled high with books, some 1200 in all, of which the Imam regarded 600 as important. They were all hand-written, certainly old, many ragged, and, were it not for the dry climate, might well have been in still worse condition. Bryce worried about rats and mice, and was sure that they ought to be transferred to a modern building. The Imam brought a couple of books out into the courtyard, and agreed to have Cécile start photography the next Sunday. She indicated to the Imam through their interpreter, their high-class Land Rover driver, which books Moktar Ould Hamidou wanted to have photographed in their entirety.

Thereafter, they went on foot to the house of Mahmoud's cousin. The dinner prepared for them consisted of thin wheat crêpes served with sheep or goat meat, some zrig, tea, and chicken served with bread. Bryce had brought along five balloons to give to the children of the family, and for the eldest boy a present, a small hand-forged knife with a sharp point, and tweezers built into the handle. Of two still older boys, one was working in Nouakchott, and the other at the University in Dakar. This boy was very interested in electrical things and

Figure 60: Cécile lecturing at the village school.

Figure 61: The Imam in his library of manuscripts.

Figure 62: The mosque at Chinguetti.

they hoped they would be able to collect some items to leave for him. His father would have liked some of their packing crates, wood being a precious rarity in those parts, but they had to explain that most of them would be needed for shipping things back to the USA, and that it would not be possible to give him anything before November.

At 9.00 p.m. they headed for the observatory, via the gîte, for their interview with the local broadcasting people, and then for their first night of observing. The interview went off well, though interrupted by a couple of messages from Mikesell, on the nightly contact with Boulder. They began with setting the alignment of the polar axis in azimuth, i.e. true north and south, and found it nearly correct, only needing minor adjustment with (! horrors !), a few blows to the support base with a sledgehammer. They then ran into trouble with the Accutrac drives, one of which was not functioning, apparently getting no power from the line. They were unable to adjust the altitude of the polar axis, because of clouds and a light dust storm, with visibility worse than during the previous year, with Bryce starting to be very worried about the success of the expedition. They quit at 2.00 a.m. and were ready for bed an hour later. A news item was that Medrud, now in Dakar awaiting the arrival of the big crowd, had made arrangements so that Mahmoud would be John Hagen's official driver in Atar. This would be great for Mahmoud, employed every day and able to spend the nights with his family.

Bryce got up late on the morning of Tuesday June 5, which brought a raging sandstorm, in spite of which, there was a school visit to the observatory. The students were passed in through the 'airlock' of the double door entrance, eight at a time, and expressed their thanks for their visit by singing two songs. Cécile went to tend her eye patient, while Bryce gave the offending Accutrac to Bob Bowie to see if he could diagnose its ailment. The makers emphasized that these items could only be repaired in their California works, but fortunately Bowie was able to replace a defective input switch. There was another school visit in the afternoon where Bryce and Cécile had been working on the darkroom, hanging black plastic sheets and fixing shelves. This time, they made all the children remove their shoes outside, and Burton vacuumed their feet before they entered the airlock, a procedure which fascinated them.

Cécile did not have supper at the gîte, having accepted an invitation to go to the house of one of the school girls. It turned out to be an all-girl party, mostly aged between 12 and 14, a few married with a lot of frank girl talk. She reported back at 8.30 p.m., by which time the others had gone back to work, mostly on the darkroom, quitting early because it was too windy and dusty for star observations. The radio men also quit early having been unable to contact Boulder.

Bryce had a bad night with intestinal trouble, and stayed in bed all morning on Wednesday June 6, while the others busied themselves with interior finishing work, painting, caulking and the like. After getting up he began planning team-work schedules, and plate-logs ready for the arrival of the others on the following Friday. There was another school visit around 5.00 p.m., with the now usual removal of shoes and vacuuming of feet. With increasing numbers of people visiting the gîte and calling on Cécile for medical treatment, or to invite them out to supper, Bryce felt that by now far too many people were coming, and that the guards needed to be more strict in the application of the rules. Although the Sun shone, the sky was milky-white, but they still told the NCAR people that they would be going to the site to work on the telescope alignment. However, the clouds came up and obscured all the stars except Arcturus, so they walked to the site in the dark to say that they would not be needing power that night. Of course, once everybody arrived, the generators would be going to 24 hours per day operation. Cécile bought three of the Mauritanian padlocks and a tea pot.

Bryce had had another bad night, and stayed late on Thursday June 7 trying to plan work schedules, plate-logs and record-keeping in time for the rehearsals for the great day. At about 11.00 a.m. he went over to the observatory carrying with him, as yet unused, the guerba, (Figure 63), which they had been given, thinking to fill it with water and put it in front of the structure 'just for style'. He had tried to devise a lever system for the precise telescope movements which would be needed at eclipse time, but finally decided that old-fashioned pushing on the tube would be better. He was full of praise for the achievements of the other two. Burton had devised a drying rack for plates, and Cécile had arranged a glass window over the eastern exhaust port, so that on the official inauguration day, June 10, visitors could file past and look inside. Otherwise the whole affair was spick and span with the floor covered with mats carefully rewoven at the edges of the trimmings.

The day started overcast, but steadily cleared. He still felt too weak for work in the afternoon, but, by supper time he went out to try to help in the telescope alignment, which was still not perfect by 2.00 a.m. They were visited during the early evening by Mr. Jacquet, (in charge of the solar pump), and by Mr. Rahael, Mr. Anezin's boss, who was in Chinguetti on an inspection trip. This caused them little inconvenience, since the work on the alignment consisted of letting the instrument track for long periods and checking its accuracy.

A more serious interruption was that when they arrived they found three of the girls who had invited Cécile to supper, waiting to press another invitation. Their presence there was in direct contravention of the Préfet's orders, but the guards could not send them away because they were there by invitation of Saidu, the chief interpreter of the NCAR group. He was a Ghanaian who was fluent in French and Arabic, as well as the English he spoke from his citizenship. It was he who had hired the laborers and a number of the guards, causing some bad feeling. Earlier in the day the Préfet had confided to Cécile that, unfortunately, the hiring of the guards had been done on a day when he had had to be in Atar, and

Figure 63: Bryce inspects a guerba (water skin)

Figure 64: Reinforcements seen off at Austin,
(l. t r.): Matzner, Evans, Mitchell, Cobb, Mikesell.

that Saidu had not consulted him. Saidu, not being a native of Chinguetti had no understanding of the various feuding factions in the village, so that it was not surprising that he made unfortunate choices, particularly since every choice of this kind was interpreted as mark of favor or disfavor. If it had been left to the Préfet in the first place, he would have installed military guards (at no cost to the expedition !). Now each military guard cost about $ 85 a month, and there would have been no need for him to issue the order prohibiting the villagers from visiting the site. Visits would then have been on a permission-only basis. As it was, the Préfet insisted on NCAR's hiring two guards of his own choice, one of them being their own night guard, and another a cut above the others in bearing and quality.

Unfortunately he would not be able to ask either of them to act as guard for the building after the eclipse until the follow-up team arrived in November, since too many factions were involved for him to risk playing favorites. (See Gerteiny, 1967 for details).

Bryce took Saidu aside and asked him to be more strict in conforming to the rules, but in spite of this, he later knocked on the door and asked if the girls could see inside. In spite of the fact that they were of school-age and should have come with the groups, Bryce assumed that Saidu had got permission from Cécile, but when she appeared she said that this was not so, and sent the girls packing. Not long afterwards Saidu was seen giving their guard hell for making such a fuss, as if he were responsible for Saidu's removal. The DeWitts tried to reassure the guard that they had complete confidence in him, but as he spoke no French the conversation had to be conducted through an interpreter. This was a man they had not seen before, who turned out to be a real nomad who had just come in from the 'brousse' for two or three days, with three camels. He had served with the French army before Mauritanian independence. Bryce's thoughts turned to the possibility of a two or three days' vacation with him after the arrival of the other members. The man had not thought of ferrying tourists, but thought it might be possible, though no details were discussed.

Again a late morning on Friday June 8, but Bryce felt fully recovered and worked on the duty schedules, but was interrupted by Cécile saying that lunch would be early. Word by radio from Nouakchott said that the DC-3 carrying all the American teams had already departed. Cécile gulped her food down, treated a baby brought in with impetigo with the usual Neosporin ointment, and joined the rest of them for a Land Rover ride to 'Chinguetti International Airport' to welcome the new arrivals.

## Reinforcements Arrive

Charles Lavern ('Chuck') Cobb, of the observational staff at McDonald Observatory, came into Austin for a final briefing with his expeditionary companions, Richard A. Matzner, a faculty member from the Relativity section of the Department of Physics, and Richard I. Mitchell of the Astronomy Department, and a farewell dinner from Charles Jenkins, Associate Director for Management of the Observatory.

On June 6 they boarded a jet at Austin, seen off by Al Mikesell and David Evans, bound for JFK New York. (Figure 64). At about 10.00 p.m. they joined some eighty other scientific passengers on a chartered PanAm Jet bound for Dakar, Senegal, on the West Coast of Equatorial Africa. They were royally fed on filet mignon for dinner, followed by a movie, which lasted until 2.30 a.m. New York time. The trip only lasted about 7 1/2 hours, so it was then time for breakfast, the Sun having risen, as it must, for those flying rapidly eastward over several time zones.

On arrival, some 20 to 25 individuals destined for Mauritania were bussed to the SU-NU-GAL hotel, where Cobb and Matzner were room mates, and Mitchell shared

Figure 65: The University of Dakar.

Figure 66: The Great Mosque at Dakar.

with Robert Tucker from California. They each got a bucket of water, the supply having been turned off for the day. The group suffered a severe language handicap, since none of them spoke French, the prevailing means of communication in this former French colony. They did manage to get themselves lunch, and to arrange a bus trip to downtown Dakar, on which they were accompanied by three of the hotel staff. After a walking tour of several shops and sidewalk markets, the employees gave the visitors a tour of the city, then estimated to have a population of half a million. This included the University of Dakar, (Figure 65), several imposing government buildings, the Grande Mosque, (Figure 66), arguably the world's largest ecclesiastical building of the century so far, as well as some poor and squalid areas, reminiscent of parts of Mexico.

On June 8 they flew on a DC-3 to Nouakchott, capital of Mauritania, where, after dealing with formalities of entry to the country, they re-boarded the aircraft for the final leg to the desert airstrip at Chinguetti, where they were met by several members of the NCAR team and the Texas advance party, who had been there for about three weeks.

The remaining passengers on the PanAm jet, who had mostly been at the N'GOR hotel in Dakar, went on that day to Nairobi preparatory to a long overland journey to the selected site near Lake Rudolph.

At Chinguetti, the new arrivals, about 27 expedition members, together with a few NCAR staff, rode Land Rovers to the gîte, where their rooms, to their pleasant surprise, had showers, wash-basins, and air-conditioning. Check-in was followed by a lunch of cold cuts and cheeses, and, from Morocco, (according to Chuck Cobb), the sweetest and best-tasting oranges they had ever eaten. It is possible to figure out pretty precisely which expeditionary individuals went where, whether to Chinguetti, Atar, or Kenya, but not so easy in the case of the NCAR staff, except for those like Medrud, Bobka, Bowie, McCullough, Morel etc. mentioned by name in

Bryce's extensive diaries, who were already at Chinguetti. Detective work on personnel lists suggests that NCAR, at all the sites in their charge, provided altogether, some 20 staff members, a truly noble effort.

At the first Texas team meeting, held after lunch, the members were brought up to date on the situation, which attested to all the hard and successful efforts of the advance team, and then walked out to inspect the site about 300 yards from the gîte. They were all ready for final testing of the operations at what was dubbed 'McDonald Observatory East'.

Next day, after a continental breakfast, it was learned that an inaugural ceremony of the whole foreign enterprise, originally scheduled for the following Sunday morning, had been moved up to that day at 8.30 a.m. They made haste to receive the dignitaries of the town, representatives of the Army, and other important people. A flag-raising ceremony was held, in which both the American and that of the Islamic Republic of Mauritania were raised. Moments later they hoisted the McDonald Observatory flag. A less than serious word should be inserted here on the subject of the, strictly unofficial, and later officially disapproved McDonald flag. Its originator was, so the senior author recollects, a visiting Chicago astronomer, Ronald Shorn, who went back to the epoch of Chicago management of the McDonald. The general field of the flag consisted, not surprisingly, of vertical stripes, more or less, of some of the spectrum colors, purple, orange, yellow, and a whole quarter devoted to blue. On the orange stripe was a prominent, five-pointed silver star. Perhaps in recognition of the facts that the color adopted by the University of Texas was orange, and that Texas was often called 'the lone-star state'.

Then, on the blue area, there was a shield, with a crest representing a thundercloud. This was quartered, the quarters representing respectively, a Coors beer can on a white ground, a formal white star on a red ground, crossed pool-cues and an eight-ball on a white ground, and finally a white representation

Figure 67: The strictly unofficial, McDonald flag.

Figure 68: The main road to Ouadane and the Richat.

of Saturn on a red ground. (Figure 67).This was all great fun, but one can see why, later, the University authorities would have none of it. At this remote time we were all rather taken with it, but what the local dignitaries might have made of it, had they realised that it contained a reference to alcohol nobody could say, and we were lucky that this never occurred either to them or us.

Details of the essential eclipse preparations may be laid aside for a few moments, to recount the experiences of the newcomers, who were clearly thrilled to bits by their ambience, lodged in the middle of the Sahara in a real Hollywood-style foreign-legion fortress—even though it did seem fortunate that its warlike capacities had never been tested in conflict.

This choice of arrangement is suggested by Chuck Cobb's excellent recollections, which read in part "The next few days were spent in preparations for the coming eclipse. We enjoyed the rooms at the gîte, and all the meals were very good. We were served lamb, steak, fish, ham, omelettes and other goodies. All meals were continental style, lunch and dinner consisting of four courses with unlimited supplies of mineral water from France, and French bread (baked locally)".

For at least some members this was a hugely enjoyable experience, but it would be entirely wrong to suggest that they were in any way other than enthusiastic, enterprising and extremely hard-working in the cause of the astronomical enterprise which had brought them there. As we shall see, they all earned the thanks of their professional world for their devoted and effective service—and all the better if there were some features of it of a most pleasurable kind.

On Friday June 15 they had their first sandstorm, not an extreme one, but too uncomfortable to allow much work outside when the sand was blowing. The temperature was not unbearable. The hottest day was 114° F, but the maximum was mostly 105-110 ° F.

Figure 69: the Richat, now deemed an erosion
feature. (NASA photograph, by permision)

Figure 70: Host of a desert excusrion.
(Photo. Charles Cobb)

On Saturday June 23, they had planned a trip to Ouadane, and what Cobb calls the meteor crater at Richat, but because of a local political conflict they decided not to go, but instead spent the night in a palm grove a few kilometers from town. Ouadane is the ultimate oasis remote from Chinguetti, attainable by well-found desert vehicles (Figure 68). The Richat, (Figure 69), is a feature well out into the desert consisting of some three or four concentric circles, the largest some 20 miles in diameter, and 2000 feet deep. Whether of meteoric or volcanic origin was not, at first, clear, though some sources such as Gerteiny asserted that it was volcanic. The NASA photograph solves the puzzle, identifying it as a wind erosion residual. At the palm grove, they were greeted by its owner, and invited to tea, (Figure 70), their first experience of this universal desert ceremony. After this affair which lasted for some hour and a half to two hours, they invited their host to eat with them, but he said that this was impossible, and that he would have been insulted had they been Mauritanian, since by local tradition it was his place to invite them for dinner. He was gracious enough to excuse their ignorance, and regretted his inability to issue an invitation since they had arrived unexpectedly and unannounced.

On another evening they watched a performance of some local male dancers, whose program was said to depict aspects of local life. The audience reclined on mats, rugs and pillows and were served the usual tea and a national drink (? zrig?) composed of goat's milk, water and sugar. This they found not disagreeable, but unlikely to be popular in the USA.

On Tuesday June 26, with most of the work done, three members of the group rented camels for the day and trekked out into the desert. After about three hours in the great expanse of sand, they met a goat-herder who lived nearby, and arranged to buy a kid for lunch. The goat looked all right to them, so it was duly slaughtered, the choice pieces put in the stomach and placed under embers, which had alreay been started. The rest of the goat, cut up, with the exception of the skin, which had been kept intact to serve as a waterbag, was put in a stew pot on the fire. While the goat was cooking, they were served tea, and then, most of the sand having been brushed off, the best cooked parts. The three camel drivers, two men and a small boy, and the goat-herd ate from the stew pot. After a few bites of his not very well cooked 'best parts ', Cobb gladly donated his to the stew-pot. After tea, they told the drivers that they needed to return, who thought them foolish to do so in the heat of the day. They insisted, but felt in the end that the advice of the drivers had been correct. They ran out of water before reaching the gîte, it was one of the hottest days they had during their stay, and even Cobb, a West Texas resident, began to suffer from sunburn. When three other Texans took a trip next day he advised them not to go so far.

Having made all the preparations for the eclipse, the team had successfully resisted the fatal temptation to try for some last minute improvements. It cannot be too strongly urged on any expeditionary or sole-chance enterprise of this sort, that when one has done one's best to perfect the preparations and finalize the routine, one should leave everything severely alone. Never, never, let any one afflicted with a sudden access of genius get his hands on the business at the last minute. Go through the routine no matter what the conditions.

This certainly justified the camel trip, and the stay in bed the next day, when there was a sandstorm, and the decision to go ahead as planned on eclipse day, even with very unpromising conditions.

We shall see how things turned out, but look first at all the preparatory work which brought the team to their perfected equipment and routine, and sent them off on their holiday taking possible itchy fingers far from the place where they could do most harm.

# Final Approach

We now turn back to Bryce DeWitt's journal, re-starting at the point where the advance group drove out to the airstrip on June 8 to meet the newcomers. According to Bryce's record, the initial after-lunch meeting did not include Gorel, who evidently had not secured a place on the local NCAR charter. At the site-inspection, the newcomers were shown all the details for checking the building after each shift and the location of all the supplies and tools. Cécile also described her activities with the schoolchildren. Bryce found that he had missed the best group visit at 10.00 a.m. that morning, consisting of little ones marshaled in military style and singing. After being shown the telescope, they sang again for Cécile and Burton, who surreptitiously turned on the tape recorder, to the astonishment and pleasure of both pupils and teachers when he played it back at the end of the visit.

The newcomers brought along to the observatory most of their scientific baggage and started unpacking it, with Cécile trying to find proper places for it to be stowed.

While this was going on, there was a knock on the door by the worker who had sewn their mats together. He was asking to see 'Madame' while at the same time, the guard was scolding him furiously in Hassaniyah. He had in his hand one of Cécile's shoes, which she had hoped some local craftsman could repair. Bryce did his best to calm things down and impressed upon them that they were not to be disturbed while at work. When Bryce went out to find Cécile, she was looking at some possible purchases brought for her inspection. The squabble stemmed from the usual source, factional rivalry among the inhabitants, which Bryce felt might be calmed if their contacts with the locals were not quite so close.

On the serious side, Bryce felt that Dick Mitchell, was too ready to lay down the law on programs according to his view, and to propose grandiose plans for research, for which they certainly could not afford the time. While looking things over, Richard Matzner discovered that his electrometer, which had been stored under the telescope base had been turned to 'OFF' when packed, and not to 'VOLTAGE OFF', so that its 9-volt batteries were completely dead, and no replacements could be found, neither from other teams, nor from the NCAR supplies.

At supper, while planning the shifts for the night and the morrow, Mitchell expressed a wish to go on night shift immediately, so Bryce put him, with Burton and himself down for this, leaving Matzner and Cobb to start next day. Cécile would be busy at the school almost the whole of that day. The two left for the observatory immediately after the meal, but Bryce had first to go to a meeting of all the team leaders that Medrud had requested. He outlined how things were to work at the gîte, and on the site, with little change from existing practice, except for a little more formality. Bill Curtis, who had come on the plane for a two-day visit, also addressed the group. He had arrived with a load of 30 T-shirts decorated with the NSF eclipse emblem, in response to a frantic appeal to Boulder by Cécile, when she learned from the Préfet that there were more children in Chinguetti than they had previously thought. Although he grumbled a bit, he was basically very friendly and spoke highly of the Texas expedition, especially of the building, designed by Al Mikesell and Charles Thompson. He wanted NCAR to adopt the same design concept for future field work. The meeting was then turned over to the expedition physician, Dr. Gillette, also a passenger on the plane. He said his role would be very simple, that he would be available at all times, though he would like to have a fixed time for sick call. Bryce asked him whether he planned to extend his services to the villagers, to which he replied that he would very much like to, and suggested that he and Cécile get together with Bill Curtis to plan a visit to the Préfet, to see how this might be set up. This, despite a negative view of the matter from the Embassy in Nouakchott, who thought it might be interpreted as an insult to the Mauritanians. Bryce remarks, 'The poor Embassy staff !'. Mrs. Ba had said to the DeWitts one evening that Ambassador Murphy was a nice man and well-liked, but that the others were stiff and ill at ease and hard to get to know. Richard McCullough had apparently shared this opinion, and said that the visits to the Ba's and to Mahmoud's family were the first to Mauritanian homes in five months in the country. His associations via the Embassy staff had been all-male affairs.

At the observatory, the night, calm and clear, the best they had ever had, was spent in perfecting the telescope alignment, which presented problems since the operation of the screw for the polar axis adjustment, crawled across the base and upset the azimuth setting when operated. By dint of changing screws they achieved perfect alignment by 4.30 a.m. with no drift off the cross-hairs in 20 minutes in all directions. Moreover the Accutrac drive was working perfectly, although they had learned that it needed two full hours of warm-up to achieve this.

Saturday, June 9, dawned, as already recorded in Chuck Cobb's diary, in a state of fine confusion with the news that the Préfet had assembled thirty town dignitaries and eight soldiers for the official opening of the eclipse effort at Chinguetti, believed, at least by the Texas group to be scheduled for the following day. Bryce scuttled out of bed, grabbed one of the kites and his camera, and dashed out to the site, where most of the team members, plus Medrud, Curtis and the official party were already assembled. The fact that there were so few townspeople assembled was possibly due to the information previously given out that the ceremonies would be on the following day. The flag-raisng duly proceeded, and was followed by guided tours of the various installations for the no more than thirty or forty people who attended. They were shown NCAR's equipment, including their refrigerators and large radio antenna, thirty feet tall with a cross-bar on top, operated by remote control. Two men invited to try the walkie-talkies were at first unsuccessful, since both kept their thumbs on the control knob and so could not hear each other. Everybody then went to the Texas building and looked into it through the glass observation window provided. After that to the Florida site, which was to be devoted to the possible detection of atmospheric pressure-changes during the eclipse.

There were almost no women in the audience and no children, but even so, Bryce flew the kite he had brought. This did attract a whole group of youngsters of pre-school age, and some older ones, presumably playing hookey. They were careful not to pass the boundary of the site, just as the Préfet had ordered, so Bryce went outside and let some of them feel how strong the pull of the string was. As Bryce remarked, this was marvelous kite-flying country, with a steady wind, great expanses and no trees. Bryce went back to bed for an hour before lunch.

Lunch time brought the arrival of Claude Morel, carrying mail, to be added to some brought the day before by Curtis, for the NCAR group. It was great to get family mail. Al Mikesell included in his written comments a list of additional items that he thought they should include in the team's activities. Mikesell also reported that the customs brokers had had two men on the look-out for the DeWitts at the Braniff and Air Afrique counters in New York on the day they passed though. They were not impressed by this since they would not have gone to the Braniff desk as transit passengers, and there was nobody at the Air Afrique counter to which they had been taken by a Braniff agent, where they remained conspicuously labeled as eclipsers for three hours.

Morel had stopped for the night at Atar, where he had seen Mahmoud and also John Hagen, who was using Mahmoud and his Land Rover for his operations. Morel said that Hagen would be visiting Chinguetti, which would give an opportunity for them to see Mahmoud again. Also the French expedition hoped to visit before eclipse time, offering a pleasant prospect of contact with other scientists. Cécile also arrived for lunch after acting as a tour guide for the inauguration and giving lectures to three classses at the school.

After a formal group meeting for assignment of duties, Bryce went to the gîte to get some sleep, reappearing at 5.30 p.m. for the last of Cécile's lectures, and to take pictures, especially of some of the younger children previously overlooked, together with tape recordings of some of their songs.

Bryce fretted that it had become impossible with pressure of all of these ancillary activities to compose a daily hand-written account of their scientific activities, and was turning to the tape-recorder instead.

The two walkie-talkies brought by the team were set up, one at the gîte and one at the site. The day crew reported progress, but Bryce still felt pressed for time. No sooner had something got done but there seemed another dozen things yet to do. Burt had mixed up a complete set of chemicals and had put the dark-room in order, but it appeared that two rinsing trays beyond the refrigeration containers used for this purpose, were required. These needed too much water, but maybe if the deionizers could be got to work, these could be used to purify local water, but if bottled Evian water were to be the only pure source, they could not use such large quantities.

Matzner worked on his radio all afternoon, in spite of a strong wind and blowing sand affecting the antenna. Chuck Cobb spent the afternoon trying to get the wooden plate-holders assembled. These were not assembled in Austin, and now presented considerable problems, but he was reluctant to consider using Johnny Floyd's metal plate-holders, so the plywood ones had become important.

The wind did not drop after sundown, and though the sky was clear, Bryce decided that too much sand would get into the building if they opened up, so that hopes of taking focus plates were foiled. They did determine the location of the optical axis with respect to the axis of the telescope tube. This was done by Mitchell and Bryce, with Matzner working on his radio trying to establish a link with Austin. He did get contacts in various parts of the world: Czechoslovakia, England and Virginia, and a weak signal from Dallas, but nothing from Austin.

They found the optical axis only half a centimeter from the tube axis, and then inserted a developed plate in the special plate-holder with a hole in the back, and prepared a lens for insertion into the hole. The first focus run, when weather permitted, would be on a bright star whose image would fall on the plate, with the emulsion acting as a ground-glass screen, so that optimum focus with the back plate in various positions could be found. The back plate had already been squared on to the telescope axis as accurately as possible with a little steel millimeter ruler. The work was all done by midnight, and they got to bed by 1.00 a.m.

This day the possibility of hiring camels for a desert jaunt had again come up, but a possible contact had not appeared. Bryce very much felt that Burton deserved a vacation, but he himself felt he could not be spared, at least not for the next five days, when they would be trying to get good photographs of the environment of the full Moon. If there were to be a vacation it would have to be in some ten days, so they should have plenty of time to locate a camel man through their acquaintances in the village.

This being Sunday June 10, thought to be the original date for the inauguration, they expected that some of the villagers and school children might show up, so eclipse buttons, cameras, kites, and so forth were made ready at the site, but nobody at all came. Mitchell did send up two kites one on a very long length of string. The other broke away and was lost.

The daily meeting report included word from Matzner of his non-success in establishing radio communication with Austin, but included the news that from the southwest corner of the building, where his antenna was mounted, he had spread out to a distance of about ten feet, over an arc of some 270°, a fan of copper wires to create a ground reflector such as the sandy and rocky soil was incapable of providing.

Burton and Cécile had gone to the mosque for their first session of photographing manuscripts, but when they got to the library Bryce found that the focal length of the lens provided for the work by Mikesell was much too short, and he could only get a portion of one page on the film. The Imam begged to terminate the session early because he was feeling rather tired, so they came away having arranged to return on Thursday. At the observatory, Burton had put black paint on the back plate of the telescope, which had previously been prepared by etching and a zinc chromate primer.

Cécile had been carving new rinsing trays from some of the surplus Styrofoam blocks that were lying around, and Burton and Matzner proposed a scheme of running water, in the form of a slow controllable trickle into the dark room to be dumped to a waste receptacle outside. This possibility depended on their being able to get some suitable tubing from NCAR—the only material Texas had was some copper tubing that had to be saved for use in connection with the generators. This was all achieved by evening with a drain installed to an outside collection bottle, this to be emptied periodically some fifty feet away. A shelf had been built for Matzner's radio : the réseau box for the photometric imprints on the eclipse plates had been nearly completed: and Cobb had assembled the plateholders, though with some difficulty.

At the end of the afternoon they had their first 'fire drill', by which was meant a run-through of the actual steps and operations to be performed at eclipse time. Bryce had drawn up a program, assigning various jobs, and taking the role of quarterback (as in American football) for himself.

# Fire-Drill

The title chosen by Bryce for this particular operation is interesting. Clearly it owes nothing in its inspiration to the operation of the pompiers dealing with a domestic conflagration. Rather it recognizes the fact that the operation is one in which every participant must play his part exactly and at a precise time, reminiscent, possibly in Bryce's mind of naval gun drills. Operations with the timing dictated by nature, and the penalty for a single mistake, total failure, while possibly not unique to astronomy, are not confined to eclipse observations, and the analogy with naval gunnery has been noted, half in jest elsewhere.

We reproduce the official eclipse schedule, in plain text, while adding in italics sufficient commentary to explain why some items are included:—

## Eclipse Schedule

T2-24 h     Start adjusting temperature to mid-eclipse. House fan blowing. Keep blank plate-holder over telescope back.

*T2 is the predicted time of so-called second contact, the moment when the Sun's disk is completely covered by the Moon, i.e. the beginning of totality.*
*The adjustments of the equipment, especially of the lens were sensitive to temperature and the whole equipment had to be thermally pre-soaked to the best temperature estimate of the probable ambient value at mid-eclipse.*

T2-4h Load plateholders. Sensitometer plates.
*All the plates likely to be exposed during the eclipse were subjected to various processes of pre-treatment. Each was*

imprinted with the grid of artificial stars enshrined in a plate produced by the instrument firm, Astromechanics, in Austin. This would have been placed in contact with each eclipse plate in turn and exposed to light so that only the grid registered on the eclipse plate. Each plate was also imprinted with a pair of so-called step-wedges located at their opposite edges out of the way of the central serious business. These small pieces of glass carried stripes of obscuring material transmitting known ratios of intensity. When developed the intensity of the images could be interpreted as a calibration of the relative intensity of incident light. This type of calibration was necessary, since photographic materials do not respond linearly to incident intensity, being reluctant to darken for the faintest lights, and saturating by the deposition of all available silver grains for intense incidence. The intention was to provide a means of changing the distribution of blackening on the developed plates into a map of the actual light intensities for which the peak incidence should be the best determination of the star's position.

T2-1. 3 h    First contact. Start telescope drive motor, but do
             not engage it to the telescope.
*The moment when the Moon's disk first impinges on the Sun. The Accutrac needed to warm up if it were to drive well.*

T2-1 h    Point telescope correctly for 1st exposure, star field.
          Load telescope Plateholder 1. Turn on photometer.
*The first exposure is to be on the comparison star-field. The photometer records sky brightness and determines choice of exposure lengths and type of plate-emulsion.*

T2-10 m    Open hatch.    Turn off house fan.
                          Turn off telescope air circulation
                          Start sector
                          Read temperatures all round
                          Engage telescope drive
                          Pick up guide star

The desired temperature having been attained as well as possible, the actual temperatures achieved are recorded. The follow-up expedition due in November to photograph the same star regions at night time, will need to reproduce these eclipse-time temperature readings by operating the installed heaters available. The "sector" mentioned is the little spinning top with opposite sectors cut out, placed before the plate to reduce direct light from the eclipsed Sun, which though vastly reduced, could scatter into the star fields and degrade the quality of the images. The ventilation is stopped to avoid currents which might transport dust to the lens or plate surfaces.

T2-5 m      Cover lens—draw dark slide

T2          Uncover lens

Predicted time for the second contact was available many months ahead, but when it came to the actual time, which might be a little off because of slightly imperfect allowance for the irregularities of the lunar limb, the time would be determined by direct observation of the complete disappearance of the solar disc, by the "quarterback" keeping count aloud of the lapse of time during totality.

T2 + 5 s    Open shutter on stars. Read photometer. Read temperatures. Note wind.
+ 35s       Close shutter. Move telescope to eclipse position·
+ 45s       Open shutter on Sun. Read photometer. Read temperatures.
+ 105s      Close shutter. Insert dark slide. Change plateholders
+ 125s      Open shutter on Sun. Read temperatures and photometer. Note wind.

| | |
|---|---|
| + 185s | Close shutter. Move to star field. |
| + 195s | Open shutter on stars. Read photometer and temperatures. |
| + 225s | Close shutter. Insert dark slide. Change plateholders. Shift to Sun. |
| + 245s | Open shutter on Sun. Read photometer and temperatures. |
| + 395s | Close shutter. Move to star field. |
| + 315s | Open shutter on stars. Read photometer and temperatures. |
| + 345s | Close shutter. Insert dark slide. Remove plateholder. |
| T 3 = | Read all temperatures. Close box of plateholders. |
| T2 + 363.7 s | Read wind data. |
| T3 + 45s | Close housing. Leave telescope untouched! Start air-conditioner and air cooler. Disengage telescope drive. Turn off telescope drive. Turn off sector. Cover telescope lens. Insert blank plateholder over telescope back Cover telescope with plastic wrap Check over all notes, temperature readings, meteorological data, everything—especially departures from procedure, individual observations of phenomena and performance. |
| T3 + 4h | Develop plates |
| T3 + 24h | Move air conditioning units into housing and secure everything for the wait. |

*In this document, which must have been the final version refined after a number of trials, T3 denotes the end of the total phase, predicted to last 363.7 seconds. The instructions lead to the production of three plates, each pre-exposed to the grid of artificial stars, and each containing, superposed, exposures on the comparison star-field, C, and the eclipse field, E, but not in the same order, being respectively CE, EC. CE.*

The provision of such a document may seem to be excessively fussy, but any astronomer with years of telescope experience, will be able to recall instances where things went wrong, in spite of what was hoped to be meticulous preparation for all the circumstances likely to be encountered. The senior author has a wealth of instances to recall, some vicarious, some, alas, personal. What has wryly been called the classical experiment of astronomy, namely carrying out the whole procedure with the cover left on the instrument, is rare, but not unknown. At a situation like an eclipse, where a controller, (quarterback), may have to count out the seconds aloud, the caller may accidentally jump a hundred if he has not got a list of the seconds written out in front of him, which he actually ticks off as they pass. Enough of this. One agonized call of "oops" is enough to send weeks and months of preparation into oblivion. Only the utmost care suffices.

## Diplomatic Rumbles

Bryce refers in his journal to some concerns which seemed to have been expressed about the special status of the Texas group. As has been remarked earlier, it had quite often been noted that the DeWitts seemed to have a special expertise in developing contacts with persons of local importance denied to lesser mortals. Bryce actually wrote "While coming in to supper in the evening, Cécile happened to get into conversation with Bill Curtis, who mentioned once again that NSF had been embarrassed by the Texas team in its relations with the Mauritanian government. He made no comment other than this, but it reflected or paralleled comments that Ron LaCount had occasionally dropped in the past. After supper Bryce felt that he would like to have this particular point cleared up, so he went to Bill and asked to speak to him for a moment. He began by saying what a wonderful job NCAR was doing, how the set-up at Chinguetti surpassed their best expectations, that everything was going superbly, that the whole NSF support program, was, to his mind, absolutely ideal, and that any fears he might have expressed in his journal of a year before, proved in the end to be completely groundless. He said that Cécile had passed on his comment about the Texas team having caused embarrassment to NSF, and that LaCount had made similar comments, but that there had been no specific details of the matter. The reply was somewhat unexpected, but as Bryce remarked, in retrospect, it made sense.

At the end of May in the previous year, Bill Curtis came to Mauritania, and on arriving in Nouakchott, went to see Mr. Ould Die, the Secretary for Tourism. This was about two weeks before the arrival of the DeWitts. The first thing that Ould Die brought up was the matter of the Texas team, asking why it was that they were making their own contacts with Mauritanian officials, not working through NSF, and attempting to bypass him. Ould Die was extremely annoyed. One of the Mauritanian officials later said that he had never seen him so angry.

Bryce then explained to Curtis, that they had been in contact with Mrs. Ould Daddah in the early spring of the previous year, she being the Mauritanian representative at the United Nations at the time, and that Cécile had gone to see her. She had been most enthusiastic about the idea of a Texas team coming to observe the eclipse, and was the first to suggest that the site be Chinguetti, a place the DeWitts had previously never heard of. It was many weeks before they knew of the existence of an NSF coordinating and logistics effort. They had assumed that all the efforts of logistics and foreign contacts would be on their own, and it was only on the eve of their departure for Africa for the site survey, that they became aware of the NSF logistics support with Ron LaCount as director, and in addition of NCAR's assignment to supply the field personnel. It was at this time that Curtis was having his interview with Ould Die. Bryce privately felt that this could possibly have accounted for the curiously cool reception of Cécile's telephone call to Mrs.Ould Daddah in Nouakchott on nearly the same date. Mrs. Ould Daddah may have described the contacts she had made with the Texans to her husband, and that this news was transmitted to Mr. Ould Die, whose annoyance at the situation may have been passed to Mrs Ould Daddah. If all this happened just before the telephone call, this may have been the mechanism leading to Mrs. Ould Daddah's suggestion that Cécile proceed through proper channels. The whole situation of the original intention for an independent Texas expedition when only a minimum of outside support seemed available, and the later establishment of the large scale official intervention, produced some awkwardness throughout the whole enterprise.

## Back To Business

Unfortunately the remaining twenty days before the eclipse were not to be devoted to a steady uninterrupted crescendo of technical achievement, reaching its desired goal with plenty to spare. The vibrant life of the oasis kept intruding in various ways, reflecting in some degree the rivalry between different Arab families, and their strict rules of hospitality, often applied in a mutually competitive style. It had even been suggested that the appointment of the Préfet, a tall handsome black man, represented a desire on the part of the authorities to have some sort of neutral referee in command. Then there were the results of Cécile's laudable desire to make some recompense to local schools, originally on behalf of what was thought going to be an isolated team from Texas, much dependent on local goodwill and assistance, for the trouble they were expecting to cause. Lastly, there was the realization that the desired objective, setting up a perfectly operating scientific institute in such difficult surroundings was an extremely ambitious one. When the enterprise was first proposed, as a separate Texas undertaking, Harlan Smith, recollecting that the 1971 effort to observe the occultation of a bright star by Jupiter, had beaten the inclemencies of weather around the world by launching no less than three separate expeditionary parties, had again proposed three separate observing locations. In retrospect, it could be realized that this was a thoroughly impracticable idea based on an incorrect analogy. The 1971 event was indeed attempted from three separate locations, but each of them was by the courtesy of the director of a finished, well-equipped existing observatory, with everything in apple-pie working order. Even if complete instrumental outfits had been available for loan or purchase, the task of installing them in undeveloped locations, would have been a triple Chinguetti, that is, quite impossible. In fact there was insufficient time to do anything but to try to get what was readily available, and, in the end, after the abandonment of this triple option it came back to the Péridier Calver telescope, and inadequate time for complete preparations. The hybrid nature of this instrument when it finally arrived in Austin, already presented problems, and its conversion and installation on a new mounting were done with only limited time until it had to be shipped. All sorts of things were left to be done on the desert site:—Paint jobs, wiring jobs, mastery of the drive mechanism and of the plate-handling routine on the telescope itself. In the hut, all sorts of minor provisions rather surprisingly,were not taken care of back in Austin, including purchasing darkroom dishes for the really quite hefty photographic plates, the construction of a light box in which the images from the grid of artificial stars (réseau) could be contact-printed on a plate to be ready for use, the construction of a box in which the eclipse plates, hopefully successfully exposed, could be returned to Austin, and so on. All this on top of getting the telescope to perform correctly, which itself included some jury-rig items, such as the mounting of the guide telescope, to be shimmed in place with bits of wood, compensating for the deficiencies of the mixed setting-circles which had survived the conversion, problems with handling plateholders and so forth.

After dealing with the delicacies of international relations, Bryce and Dick Mitchell went to the observatory to try to establish optimum focal settings and other instrumental matters. Mitchell had already provided a manual of treatment for the astrographic lens emphasizing several very delicate matters of handling, already addressed when it was installed on the telescope. This included, for example, an injunction not to talk over the lens, because the coating put on in Austin was liable to damage by saliva. The two doublets of the Petzval lens were held in position by a metal structure, which was liable to size change when exposed to a wide range of ambient temperatures, as might easily be the case in the Sahara. The careful temperature regulation of the hut, and the control over the instrument temperatures by the installation of thermistors, were intended to counteract such effects. In his manual Mitchell had insisted that the best focus should be found by investigations at the plateholder end of the telescope, and not by adjusting the temperatures, though he insisted that even when focus plates were to be taken at different settings of the plate position, the relevant temperatures should be recorded.

The observers found a pretty good focal position just by eye inspection of the images of Arcturus, but then ran into difficulties caused by irregularities in the engraved setting circles of the telescope, and the verniers used for positional work. They also ran into a new source of contamination. Cécile had been cutting rinsing trays out of Styrofoam blocks, working outside the building, but it was found that minute fragments had been carried inside and were getting onto important places such as the optical surfaces. This showed that it was necessary to be fastidiously clean at all times.

Burton and Mitchell came on later that night and

succeeded in obtaining two focus plates, both on the IIIa-O emulsion—the faster, and very slightly grainier of the two plate-emulsions available. The images, though not perfect, seemed rather good.

However, in the morning, (Monday June 11), the affairs of the village suddenly intruded. There had been the problem of the apparent switch of inauguration day to the previous Saturday, and now, it seemed, the notion had developed among the teachers that the whole school could come on the Monday. They asked Saidu, who referred them to the Préfet, who said it would be okay with everybody, and told Saidu, but he waited until Monday morning to tell Cécile that they were all coming. She, at first, said it wasn't possible and that he should go to Medrud. So Saidu did, who also said it wasn't possible, and that Saidu should tell the teachers, but he returned saying that the teachers had already arrived. Cécile had to drop everything to act as a tour guide, Mitchell pounded on Bryce's door to get balloons, the children got excited, and broke their previous neat ranks into barely controllable groups, which were brought into some sort of order by demonstrations of the walkie-talkies.

After lunch, the team had another fire-drill and got plate-changing times down to sometimes to less than 20 seconds. After a brief meeting of team leaders called by Medrud, including deciding what to do about tipping the staff at the gîte, Burton and Cobb went to work on the focus, and Matzner to operate his radio, and made a successful contact with someone in Austin, and so to Al Mikesell. They complained about the trays and lack of space in the dark room. The others mastered the problem of telescope pointing but had trouble with the focus plates, including contamination in the dark room from droplets of some chemical. Bryce decided that the system of operations should be more rigorous, with never more than two people in the observatory at night, no distraction from radio communication during telescope work, that unless absolutely necessary, all construction jobs should be done outside, and that control of cleaning up and tool storage should be more rigorous.

At next day's (Tuesday June 12) group meeting, a discussion about darkroom procedures elicited the fact that the Evian water in bottles from France, supplied in large quantities by NCAR, might be inadequate for careful developing jobs. Thus, getting the team's deionizing equipment into operation became a high priority. There might be some help from NCAR which was getting its own deionizing equipment into operation after the delayed receipt of some needed parts.

The afternoon fire drill was a great success, in which there were now to be fewer responses to Bryce's commands, but would include a time signal to let everybody know how much had elapsed.

Chuck Cobb was having serious difficulties with the plywood plateholders because of a defect in their basic design. They were too thick to go into the plateholder basket at the rear of the telescope, so efforts were made to use some metal pieces scrounged up round the village for the manufacture of replacements. Bryce and Chuck spent the night trying to set up the hour-angle circle, but Bryce was annoyed with himself for making a mistake in understanding how his substitute circle should work.

Bryce's diary seems to have skipped a day, since the next entry is also labeled Tuesday June 12, and records a string of essential housekeeping chores, such as starting the installation of the thermistors on the telescope, making up a batch of chemicals for the coming night and improving darkroom facilities and equipment. Later Burton and Matzner worked on the exposure box for the réseau and photometric step-wedges. At fire-drill they tried out two different scripts one in which three III-O plates were exposed during the eclipse, in the other two II-O and two IIa-O's.

They returned for supper to find that Alassane Sy had arrived, having driven up from Nouakchott, all in one day, in the same Land Rover that carried Frank Oral of the Hawaii team, who had missed his charter connection in New York, and so came on his own. He had been assigned the role of chief scientist or leader for all the combined teams, and would make the decisions if there were any scientific conflicts.

Alassane Sy, (or Gorel as he was called by his family), went out with Cécile after supper to see the observatory, (Figure 71), while Bryce tried to improve his arrangements for hour-angle setting, helped by Chuck Cobb. Bryce and Burton took three plates during the night, and were dismayed to find large tracking errors, which they finally ascribed to the Accutrac which had not been running for a full two hours. Some later plates confirmed this. They also found, when trying to economize on the rinse water by taking the plates from the fixer to the hypo clearing bath without a rinse, that the emulsion had a series of specks due to tiny bubbles, as well as a brown stain.

The following days were occupied with these demanding small tasks, getting settings, such as the focus, fixed at their correct positions, learning the idiosyncrasies of some of the working devices, perfecting the conditions for processing plates, and steadily improving the working conditions in the hut. Occasionally crises of various kinds would arise. On Wednesday 13 June, Dick Mitchell was trying to regulate the evaporative cooler which worked irregularly because of the unpredictability of the valve controlling the flow of water into the lower reservoir, while Chuck worked on getting the deionizers operating. Medrud took Bryce aside after lunch, and said that the Texas team was causing a water crisis by their decision to rely on Evian water and not on deionized local water. They were, in fact, getting through 2 1/2 cases a day, a rate much greater than NCAR had expected. He asked them to cut back on their consumption, at least for the present. This created a special problem since the team were planning a large series of photographs of the full Moon, because of the similarity of the light intensity and its distribution, to that expected at the eclipse.

Figure 71: Alassane Sy in desert rig.

Bryce was very contrite about confronting NCAR with demands not mentioned in their original request. He also felt badly in having to ask NCAR to make up more concrete blocks for the telescope support beyond those originally scheduled. It was taking an enormous amount of water to fill up the trays used for the large foot-square plates, which is one reason why they were working on manufacturing smaller trays. The NCAR equipment was not a deionizer, but a purifier, and it became urgent to get their own working, not on Evian water, but directly on Chinguetti well water.

They later had a fire-drill using the two alternative scripts. For the first one, two of the team were still asleep after night work, so not only was Gorel initiated, but also Dr. Gilette, the medico, as a volunteer.

In the evening, Mr. Shurtleff, deputy chief of mission at the US Embassy in Nouakchott, whom they had met in May, came with his wife, and Paul Inskeep, whom they had met at Niamey in Niger. The latter had finished his tour and was on his way home. As Cécile came into supper, she was called out by a messenger, and going to the gate of the gîte, found the Préfet, who asked her to deliver a message to Mr. Shurtleff. During the day, the latter had asked the Préfet to have tea with him, and the Préfet had come to say that he could not come because of an elders' meeting. He did add, that it would not have been proper for him to have accepted, since it was he, rather than Mr. Shurtleff, who should issue a tea invitation. Cécile was embarrassed at having to relay this information to Shurtleff, that this should really have been done by Claude Morel, but it would seem that the Préfet felt more at home with Cécile than with any other person on the American team.

That night was devoted to a series of test exposures taken by Richard Matzner, designed by Bryce to reveal the sources of tracking errors, including pre-loading of the drive both against the direction of motion and in the same sense.

Thursday June 14, Bryce was up early and went with Cécile and Gorel to the mosque, where they found the son of the Imam, who took them into the library and brought out the book, of which Moktar Ould Hamidou had asked them to try to make a complete photographic copy. Bryce was able to expose a little over 100 frames, when the son asked to terminate the session for that day. Plans were to meet each Sunday and Thursday. The Imam's son taught Arabic at the local school and was free on those days.

The five plates from the previous night were pronounced successful, and were supplemented by two of the Moon, obtained by cutting through part of the roof. This was no problem since the designers had thought of this possibility and the holes could easily be repaired by replacing the appropriate sections.

The importance of the lunar photographs was that they could provide almost a dress rehearsal for the eclipse. This was now the full Moon two weeks before the crucial new Moon. The brightness of the corona at a total eclipse was often stated to be comparable with that of the full Moon, so that it would be of the greatest interest to see how the special coatings performed and what sort of stars could be glimpsed close to the Moon. Mitchell was busy with charts trying to identify those that showed on the plates and to determine their magnitudes.

Before lunch it looked like an impending sandstorm so the building was battened down and sealed. There was otherwise good progress. Chuck had got the deionizer working and produced five or six liters of water. Fifty cases of Evian had been ordered and were due to arrive on Saturday June 16. This ought to solve the water problem, and it was even considered passing some of the Evian water through the deionizer for use in processing the eclipse plates. The team was thinking more and more of using Johnny Floyd's metal plateholders, since the crack changeover team, Cobb and Jones, had been particularly good in the fire-drills.

Friday 15 June was a bad day. The night crew took only one plate before feeling ill. Then the wind got up and began to tear the tent apart so that very little outdoor work was done. Several members were feeling rather unwell and tired, and considered the idea of a vacation on the following Tuesday or Wednesday. Possible ideas included renting one or two Land Rovers and visiting Ouadane, the outermost oasis, for which good chauffeurs and experienced guides would be necessary, or they could rise very early and walk to a palm grove some ten kilometers away, and spend the day, perhaps even the night, just relaxing. Or they could try to rent some camels and ride to some palm grove or other. It was proposed to consult the Préfet.

At the observatory they fixed up some shelves to hold the equipment now resting on the floor round the base of the telescope. They also improved the wiring by installing change-over switches for use if a quick change of apparatus should become necessary at the eclipse.

During the afternoon, the Préfet's assistant came to invite the DeWitts, Gorel and Dr. Gilette to supper at his employer's house. By the time they were ready to go, the wind had died down, and though it was no longer the season for real sandstorms, there was a considerable amount of dust in the air, and the rising Moon was only visible at a fairly high altitude. At the Préfet's house, they had supper on the roof, on mattresses as usual. Bryce had put on his Mauritanian boubou for the occasion, the food was excellent, and, at the end, Cécile was fast asleep on her mattress. Just as they were leaving to go to this supper, Gorel came out of his room wearing his boubou, to be taken by the guard for a local intruder, who ran to chase him away, until he saw who it was. The guard said "Oh you're wearing a boubou." Gorel remarked, "They took me for a Mauritanian."

On Saturday June 16th there was a strong wind which stopped the night crew from doing any outside work, but they had perfected the sensitometer routine needed for the eclipse plates. Richard Matzner had got sick in the night, but had been put back on his feet by the doctor. Nearly everybody had had, at least a touch, of intestinal upset, but work had gone well forward on improving wiring, and other tasks and the fire-drills had become more and more expert. This was the evening when everybody went to a performance of black singers and dancers held in a courtyard in the old part of Chinguetti, far enough away from the gîte so that the noisy generators, a sharp contrast to the quiet of the previous year, were inaudible.

The full Moon shone down through a windy and dusty sky on the audience reclining on mattresses, who were served tea and a special drink prepared from the baobab tree. After the performance, the DeWitts, Burton, Chuck and a couple of men from another team, went for a walk accompanied by some four or five young boys. In the absence of the heat, it was most pleasant in the dunes in the moonlight, climbing them and running footraces with the boys, though soft spots occurred without warning and made the runner fall down.

The boys took them to a deep hole in the ground, equipped with a pulley, a rope, and a bucket, which, they explained, was not a well, but was used to get banco used for surface material and mortar in local buildings. They were told that there was a source of this mud at the bottom of the hole, that there were two or three such holes in the area, and that, if one went down (they offered to take the DeWitts), one could wander for a great distance through a natural cavern and come upon the buried remains of an old town with courtyards, completely underground. One could even reach almost to the gîte by this underground route. After teaching the boys, at their request, some English phrases, which they picked up rapidly, the DeWitts decided against sleeping out in the dunes, because the breeze always blew some sand around one's head.

On Sunday 17 June, they were at the library with the Imam and his son, and one of the school teachers. They continued with the photography of the book suggested by Moktar Ould Hamidou, which turned out to be rather long, some five hundred sheets, hand-written on both sides, but after they had taken three rolls of film, that is, close to 100 frames, the Imam's son terminated the session. They had done about half the book, and wished they could do more in a single session, but it was understood that they make a contribution to the mosque every time they came. It appeared that these books belonged to private individuals, but were left in the care of the Imam. Cécile, in spite of the fact that she, as an infidel, was not allowed to touch any of the manuscripts, was making an attempt through one of the school teachers to try to have up to four sessions in a single day. Without this, they would not be able to make any sort of dent in the library holdings. As it was, they hoped to complete the selected book, regarded by Hamidou as the oldest and most precious in the collection, and then a complete copy of a much shorter, seventeenth-century, work which mentioned eclipses and other astronomical phenomena.

After this, Cécile went with Dr. Gilette to interpret at some house calls he was making, and this took much longer than expected, because the family presented not only the original patient, but brought in all sorts of additional members.

The diary entry for this day is a very long one, and goes on to recount such detailed preparations as providing Bryce with a clip-board for the program script and an electric clock to be started at eclipse second contact, to control the verbal instructions. Then came a session devoted to installing hour angle circles adequate for finding a bright star in the visual finder. After supper, Medrud had a meeting of team leaders covering the question of baggage limitations for passengers on the charter aircraft after the eclipse. This was to be a DC-3, to carry three extra people and to be taking off from an altitude of 1600 feet at a hot time of day, from a shorter runway than the one at Nouakchott. Meanwhile Burton Jones and Richard Matzner had stayed at the observatory taking a number of sunset plates and correlating their densities with the sky-photometer readings. Burton and Bryce then went into a series of operations designed to establish fiducial marks allowing reasonably accurate Hour Angle settings, and then a study of the declination settings, especially in view of the fact that the declination would have to be changed back and forth during the eclipse and to come immediately to the desired sky positions. They concluded that these changes would have to be made more gently than those made up to then in the fire-drills. Lastly they worked on the problem of arranging the guide telescope so that it would be pointed at the bright star, epsilon Geminorum, when the main telescope was pointed at the Sun during the eclipse. This required dismantling the whole thing and cutting new V-notches in the blocks of wood that served as positioning brackets. They did not have time to finish this by 3.00 a.m., when some clouds came, and there were no longer any bright stars to aim at. They finished this marathon session by thoroughly cleaning the observatory and the darkroom, the latter especially difficult with only a trickle of water from the deionizer.

They slept late next morning, Monday 18 June, while others did a considerable amount of work round the observatory. Gorel and Cécile cleaned up the NCAR tent, where supplies not needed inside the hut were kept, and staked out the holes where the anchors to retain the building were to be buried for the long wait until November. Matzner spent the morning, modifying his radio so that he could communicate with Atar. This was done at NCAR's request, since, until then, John Hagen, who was there, could only communicate through the French, to their Embassy in Nouakchott, who would then relay the messages to the American Embassy there. If direct contact between NCAR and Atar could be established, messages could go direct to Boulder or to the American Embassy in Nouakchott. Mitchell and Cobb spent the morning working on the roof of the hut, sealing up the various ports, and caulking.

While the daily meeting had cause to feel considerable satisfaction at what had been accomplished, it had become urgent to figure out exactly what had to be returned to the US after the eclipse, and what would be left in Mauritania. Medrud needed all this information so that he could draft the necessary customs forms and determine what weight allowance could be made for the charter flight back to Dakar. Bryce reckoned that if the eclipse was successful they would have to transport about 100 pounds of material in excess of personal luggage, including photographic plates of various kinds. If it were a failure the weight would be reduced to about 50 pounds, mainly the lens, which would have to be hand-carried as it was on the outward journey.

Cécile had a long and complicated meeting with the school teachers concerning handing out expedition T-shirts at the school prize distribution. The complications arose because NCAR had supplied 36 special T-shirts, of three different colors, that there were 7 classes and 8 eclipse teams. The original idea was to have one classroom assigned to each team, but the teachers elevated this into a much more complicated arrangement involving a great deal of protocol. The special shirts from NCAR, being fewer in number, were to be given to the best students, while Medrud, as a leader of the whole thing, would give a special shirt to the best student in the oldest class. Various team members and town dignitaries would give shirts to appropriate students. There were 305 students, and the teachers were planning to have an appropriate donor for each of them. For each, the teacher would read out two names, that of the donor, and then that of the student. The protocol for this matter appeared to be very important for the townspeople involved.

Thoughts had turned to the possibility of a vacation day for the Texas team, and Gorel had found that Mr. Anezin offered the free loan of his Land Rover for a trip to Ouadane. This was a most friendly gesture, and they decided to go on the following Friday and Saturday. Gorel went looking for a guide to take them there.

Mr. Jacquet agreed to do some of the expedition photography using the rolls of film given by the National Geographic Society. He was a great amateur photographer who loved to photograph around the village. If they got the 16 mm film to make a sequence for the BBC, as had been suggested in one of Mikesell's communications, he would have been happy to do some of the filming.

That evening, Burton and Bryce worked on the alignment of the guide telescope, though they found that having achieved the perfect alignment, the slightest pressure on the guide telescope caused a shift. They had thought of using the Questar as a supplementary guide telescope, but it proved impossible since that instrument had to be returned to the USA immediately after the eclipse, and could not be left for the November party. After that, they exposed two plates, one a III-O, the other a IIa-O, on a sky field including Arcturus, at almost the same altitude as the eclipse field, in a sky reckoned average, that is, according to figures from the New Mexico team, having possibly as much as a 50% extinction, to try to determine the limiting star magnitudes which might be reached in a given exposure time. As Bryce remarked, the sky had never been as clear as during their visit the year before. Finally they did a focus check, reckoned nearly perfect, though when the existing filter, which had a scratch, was replaced by the one to be used at the eclipse, they might have to do a repeat.

Next day this sort of work continued. Bryce and Gorel made notches at five places round the edges of the plateholders, at corresponding points on the telescope back plate. These notches were located at the positions of five prominent stars in the eclipse field which fell very close to the edges of the plate, the idea being that, even if they did not position the plates perfectly at eclipse time, they would not risk losing these important stars.

At the same time, Mitchell was installing the sector, on an arm just in front of the plate surface, carefully positioned so as not to cover any important stars. It would also be necessary to mount little metal plates on the backplate to cover those places where the step-wedge exposures and fly-spankers would fall.

Burton ran into some difficulty in developing the plates taken the night before, with spoiling of the emulsion in one case and the development of little bubbles in the other. The cause was evidently a disparity of temperature between the rinse and developer solutions. In order to keep the plateholders perfectly dust free, each was taken apart, cleaned and put in a plastic bag, from which it would only be removed for night work. Except on eclipse day itself, when development would have to be done in the afternoon, from hence forward all developing was to be done at night, with only the individual doing the developing allowed in the building.

In his diary, Bryce recalls several minor annoyances. The float valve on the cooler refused to work automatically, so that to keep the place cool required continual attention. The air-conditioner gave trouble with the cold air vent that Bryce had had to manufacture when it was installed on top of a box against the west wall, instead of on top of the roof. The trouble was that the unit was designed for an automobile camper, to be placed on a flat roof, which it could not be on the observatory building. The home-made vent made of cardboard and tape, often leaked cold air on to the thermostat wire located in the return air vent, and this made the air-conditioner recycle much too rapidly. The compressor then failed to operate and drew great surges of current, which though affecting the NCAR generator only slightly, did cause the lights to flicker all over the camp. Bryce was sure their own generator would not be able to take a sudden load of 40 amps, and though the problem had been kept at bay by patching up the air vent, it was something they had to keep an eye on.

That evening the DeWitts and Burton went for a walk to the northeast through the town. The stars were out more brilliantly than they had been for some time, and it was to be hoped if the trend continued, that the skies would be as good, or nearly so, as they had been a year before. After a while they came upon a courtyard where actors were putting on a play accompanied by tom-toms. They did not stay because the asking price for admission was a thousand francs per person, (roughly $ 5), which they thought was too expensive and showed that the village was now geared up to make good money from tourists. To modern eyes this mostly goes to show how values have changed in the last three decades.

They were soon joined by two little boys, the elder being the same one who had invited them to tea in the afternoon of the day spent in Chinguetti a year previously. They went first to his house, and spent some time talking to his sister, who gave Cécile a woven key chain, with a gift of a Chinguetti eclipse button that Bryce was wearing, in response. They then went to the house of the younger boy, (their fathers were brothers), and sat down outside under the stars. While they were there, a small caravan of five camels arrived, and they took the opportunity of enquiring about renting camels to take them somewhere for a vacation. The discussion of price lasted a very long time, and while it was going on a very bright meteor, far brighter than any of them had ever seen before, flashed across the sky. It was to be hoped that this was a favorable omen for the eclipse.

They got back to the gîte about 11.30 p.m., feeling that it was a pity that no photographic work was being done on so good a night. This had to be forgone because Dick Mitchell had installed the sector in the afternoon, held by C clamps for 24 hours for the glue to dry.

Next day was Wednesday 20 June, and Cécile, having taken some medicine prescribed by Dr. Gilette, was feeling much better, so went off to the school, taking all the T-shirts with her. These she gave to the teachers, who were preparing for the *distribution des prix* due next Monday. After that she went out to the observatory and oversaw the installation of the large tent stakes. These were set in concrete in holes dug by the laborers about eight feet out from the four corners of the building. W hen the building was closed after the eclipse, cables from the roof were to be attached to secure it against strong winds. None of them believed it would blow away even if not tied down, as it was very sturdy and much praised, and the skirts were already piled with earth and rocks, some quite large, to a height of a foot or so all round.

Bryce spent the morning computing eclipse times as accurately as possible using Naval Observatory tables, but had asked via radio to Boulder and Al Mikesell, for best estimates of contact times from the Naval Observatory. They also set the telescope as accurately as possible to the computed eclipse position to make sure that it would then clear the roof opening. It was nearer than originally thought but there was no question that it would clear adequately. After working on the roof with Gorel he computed a starting position for the fire-drill of that night, due to start shortly before 10 p.m. The drive would be started at 9.45 corresponding to $T_2$ less ten minutes, so the notional eclipse would begin at five minutes before the hour. He was happy to find that a guide star, delta Herculis, similar in magnitude to epsilon Geminorum, would be available as a stand-in. This was to be a fully-fledged test run, with the sector operating and actual plates exposed. Cécile was to be shutter man on the roof, with the shutter attached to her arm, so that there would be no risk of its blowing away. Gorel would also be up there to remove the aluminium foil blanket that covered the end of the lens.

They had ordered 50 cases of Evian water from Nouakchott, but the truck that was to bring them had broken down, and they had been forced to use Vittel as a substitute, which was much less pure. The local well water was far too dirty, so they decided to use Vittel passed through the deionizer. At the same time they felt that they should make an effort to produce distilled water. Chuck Cobb and Richard Matzner got a still operating using a jerry-can on the stove using butane, which heated better, but produced more soot than when gasoline was used. The initial success proved that it was possible to distill water, but they did not know how much butane had to be used to produce a significant quantity.

Dick Mitchell finished adding the little cover-pieces to the plate-holders to shield the step wedges and fly-spankers during the eclipse.

Richard Matzner worked on the photometer. They had changed the plan for its use. Instead of mounting it on the telescope, pointed at a part of the sky some distance from the Sun, it would now be aimed independently and it would be up to Matzner to decide whether to use only III-O plates or to insert some IIa-O into the series. The plate changers, Burton and Chuck worked so rapidly that Bryce had considered the possibility of getting an extra plate in, particularly if the sky should be of such a quality as to allow the use of the IIa-O plates. New and better darkroom procedures had been instituted and the last set of focus plates had been a great success. Chuck also improved the working of Johnny Floyd's metal plate-holders. The only unhappy feature of the whole scheme was the mounting of the guide telescope whose wooden brackets allowed too much play. A temporary measure was to put on a sign, saying "Do not touch".

Cécile came in late for supper, having gone at five o'clock to the house of the school teacher who was usually present at photography sessions. The nephew of the Imam, who held the manuscripts and turned the pages, was also a teacher, but he spoke no French, and the other man came along as an interpreter. He had previously indicated that he could arrange it so that more pictures could be taken in a single session, but matters turned out to be more complicated. The family who owned the manuscript being photographed had refused this on many previous occasions, which is why

Moktar Ould Hamidou was especially interested in it. The refusal even extended to a Libyan delegation, which had wanted to photograph and make a complete survey of the library, and had offered a lot of money. Now the Texans were being allowed to photograph some part of this manuscript, but there was reluctance to let them photograph it completely. That explained why the sessions were short and why a thousand Mauritanian francs was paid for each one. He said he thought he could work it out all right. The Imam's nephew would take the scientific manuscript to his own house where it could be photographed without the knowledge of the Imam. This left the DeWitts uncertain. When Gorel was told of this, he said that it seemed to be an attempt to see how much they would pay for photographing the library. It was decided to emphasize to the Imam's nephew the following :—That they had strict budget limits: that they had no personal interest in the library as scholars : that Mr Moktar Ould Hamidou had expressed an interest in the expedition and told them of the library, and they were merely trying to return a kindness by photographing a particular manuscript : that scholars at Princeton, who did not even know of the existence of the library, would like to have some idea of its contents. It would not be necessary to photograph any of the books to achieve this since, Moktar Ould Hamidou could supply much of the information: if further photography were impossible they could just stop.

The full fire-drill, using actual plates, proposed for that night, was reduced to a normal dry run because the sky clouded up. Afterwards they spent a long, and not entirely successful, time, trying to ensure that when the telescope was shifted back and forth from the eclipse position to the comparison field, it returned accurately to its desired position. This was particularly important because of the existence of three stars close to the Sun, which could be blotted out by the sector if the re-setting were inaccurate. Bryce considered attaching a wooden lever to the telescope frame to allow it to be moved very gently within the ten-second interval specified in the drill schedule, allowing the operator to keep his eyes firmly on the declination stops. He went to bed with this problem still on his mind.

Next morning, 21 June, the night crew reported that no sky plates were exposed because of the clouds, but had prepared three plates by putting the réseau and the step-wedges and fly-spankers images on them. They could be used for obtaining limiting magnitudes later.

He and Cécile went to the library for another photography session, and finished the book. However he did not believe that they had restarted at the point they had reached at the previous session, and that the Imam's nephew was deliberately leaving out a number of pages, possibly chosen at random. In spite of this, the session was pleasant, and the Imam cordial, especially to Cécile, when they said good-bye, which was unusual, considering that an Imam is not supposed to take notice of women. He paid her a compliment, through the interpreter promising to make her chief of the town should she come back.

At the observatory, they found Chuck hard at work distilling well water. During the morning he had produced a little over two liters, but it was still not clear how many tanks of butane it would take to produce a given quantity. A dust cover had been made to protect the plate drying rack, as part of a drive to get the darkroom as dust-free and clean as possible. The concrete to complete the system of tie-down cables had been poured. The wooden lever for the declination change was in place. It was proposed to use the Questar as an additional guide telescope, and to leave it until November in spite of the Astronomy Department request to have it back in the summer. The whole guiding problem turned on the need to line up on epsilon Geminorum at eclipse time. The declination stop was modified by putting in a threaded bolt against which there could be a firm positive pressure at a pre-determined exact position. If the sky cooperated, the night's fire-drill would include at least one real plate.

At the library, with the nephew of the Imam and the teacher who spoke French, Cécile completed a 17th century scientific manuscript, together with pages carrying tables of which the calligraphy was especially fine. Color photographs were made of those painted in color.(Figures 72,73,74).

The manuscript that contained references to old eclipses and comets appeared to be a short astronomical treatise. It turned out that the Imam's nephew, whose name was Sidi Ould Habott, was the son of the owner of the big manuscript that they had photographed. The Habottt family owned most, or possibly all, of the manuscripts in the library.

Dick Mitchell replaced the sector motor with another which would certainly not obliterate those three stars close to the eclipsed Sun, if indeed they could be seen against the glare. After supper they had a full-scale eclipse drill, with Gorel manning the declination changes, which he did extremely well. Chuck and Cécile were on night duty, and the sky being excellent they made three limiting magnitude exposures, developed the plates and left them to dry.

Next morning, Friday June 22, the general feeling of optimism was crushed by the discovery that the plates were terribly contaminated, with all kinds of spots making them thoroughly filthy. They immediately began to try to track this problem down with a program of thorough cleansing of everything. However, before changing anything, they decided to test a fresh III-O plate, to be taken out of its box, developed immediately with specially prepared fresh developer using deionized Vittel water. The shipment of Evian water from Nouakchott had still not arrived and the water problem was becoming critical. A second plate was to be exposed to the sensitometer box during the day in a darkened building, placed in a plateholder on the telescope, and the dark slide pulled

Figure 72: One of the library's calligraphic treasures.

Figure 73: Another treasure.

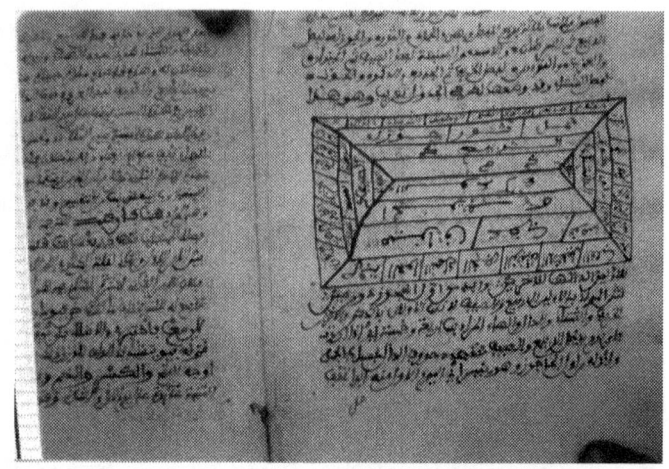

Figure 74: Could this be an astronomical text?

Figure 75: A solemn occasion,
the *distribution des prix*.

as it would be in a regular run. It was hoped that this might pinpoint the source of the trouble, and other sequences were considered if this was unsuccessful. All the plateholders were carefully washed and the inside of the telescope tube cleaned by wiping it with damp rags.

All the test plates came out as clean as could be, and no contamination could be detected, even when the brush of the vacuum cleaner was rubbed directly against the plate. The source might have been something on Chuck Cobb's clothing, but he was off sick with the sort of intestinal disorder which seemed to be going round the village just then.

After a thorough round of cleaning, they had another look at the guide telescope situation. The Questar was now mounted on the cradle of the telescope,(See Figure 56), and they tried to align it so that when it pointed at epsilon Geminorum, the main telescope would be pointed at the eclipsed Sun. Although easy to adjust, the Questar turned out not to be stable enough, so they went back to centering the five-inch finder as rigidly as possible. Burton and Gorel were to spend a couple more hours repeating the exposures that Chuck and Cécile had attempted the previous night. Tomorrow was to be the vacation with a Land Rover lent by Mr. Anezin, together with a driver and guide to Ouadane and the Richat, sleeping out somewhere.

They awoke on Saturday 23 June to a real crisis, all ready at 5.30 a.m. for their vacation trip, but Gorel and Burton had had the same trouble with the plates as Cécile and Chuck. They decided to stay in Chinguetti at least until after lunch, while they repeated all their tests in search of the gremlins.

Richard Matzner got on the radio and was able to talk to Al Mikesell directly, but he was just as mystified as all the rest. They had some suspicions of the sector creating a field of static electricity when running, but test plates, one with the sector running, the other not, were inconclusive. They also suspected a problem of plate temperatures. So far, only the nightime plates taken with the sector running had been bad.

The trip to Ouadane had to be called off. They were all ready in the Land Rover with the driver, waiting for the guide to show up. They eventually sent one of the workers at the gîte to fetch the guide, but he returned, saying that the guide wanted them to go the adjutant, the chief of the local militia, to ask for permission. This seemed strange, and Bryce and Gorel went to see him, even though they had to rouse him from his siesta nap. It became clear that he was very angry, not with them, but with a local administrative hassle. The Préfet had assigned to them as a guide one of the soldiers, and because this man would be on active duty during the trip, they would not have to pay for his services. However the Préfet had done this while the adjutant was out of town, and he was much annoyed that the assignment had not been made through him. He would be glad to assign a different guide, but he could not let the Préfet's man go with them. They immediately said that they did not wish to become involved in an internal village matter and would call off the trip, and left abruptly.

Bryce was very disappointed, though the rest of the gang took it stoically. He thought they might have offended the adjutant, which would be very important, since it would be the Préfet, acting through the adjutant who would assign the guards for the building during the long wait until November.

So, of course, there was no vacation, and Bryce was particularly concerned for Burton Jones who had been, perhaps, his strongest support, a faithful worker day in and day out since their arrival. As an alternative, Bryce began to think again about the possibility of hiring camels. This had been done perfectly well through a little boy until the question of price came up, and this they would have to settle with the Préfet. But he was procrastinating, saying that the Office of Tourism would have to decide what were fair rates for tourists for all sorts of things. He had let it be known that they were not to be taken on camels without a guide authorized by him, presumably out of goodwill and a sense of responsibility for them. "But Chinguetti is no longer the free and easy-going village of old. We cannot make our own arrangements. We are trapped". So said Bryce, who felt so frustrated that he said to Cécile, "To hell with the Préfet: let's go see see Moulaye Ahmed Ould Sidi Ali, and see if we can go tonight to his oasis. Mr. Anezin's car and driver are still available to us if we want them".

Bryce's frustration is understandable, but in fairness to the Préfet, it must be remembered that this small—in population—country, through no fault of its own, had suddenly been called upon to cope with thousands of visitors, some perhaps with the sort of irresponsibility which seems to infect many travelers when away from their home countries, most, in this case with a serious scientific goal in mind, but still a huge alien injection, capable, often innocently, of causing trouble, or of getting into it and having to be rescued from an environment which certainly could be dangerous to the inexperienced.

They found Moulaye Ahmed at length, at the school, rehearsing for a play to be performed next day in honor of the Minister of Education. He agreed to take them to his 'palmeraie', so they all packed into the Land Rover ready to go, only to meet with another disappointment. The one to which they were driven was not the one that he had first spoken of, which was about 10 kilometers away, but a much nearer one, and less pleasant, only about 2 kilometers away and even still within sight of their building. There had hardly been need of the Land Rover to get there. The explanation seemed to be that Moulaye Ahmed's friend, the little boy, who belonged to the family of Malmoud Ould Amar Cheine, whom they had approached about camels, was earlier to have obtained some to take them to the other palmeraie. Moulaye Ahmed did not want to short circuit the possibility of his friend's making some money by taking them to the other palmeraie. He made an excuse to the driver saying that the sand dunes were too difficult for the Land Rover, which seemed doubtful since his palmeraie was on the way to Ouadane over the dunes.

The term, 'Moulaye Ahmed's palmeraie',really meant a palm grove belonging to some member of his rather extensive family. The man who welcomed them was pleasant enough, served them the customary tea, and they spent the evening lying on mats spread on the ground.While they were still having tea another gentleman came to visit. He could speak very little French, but addressed them in Hassaniyah which Moulaye Ahmed translated. The old gentleman said that there would be no eclipse and that he would pray that there should not be one. He chanted an appropriate prayer which Bryce recorded on tape and, after he had talked for some time, Bryce played it back to general astonishment and amusement.

Cécile and Bryce wandered off to the top of a nearby dune and spread out their ground cover, though sleep was not easy because of sand blowing in their faces. It wasn't the vacation they hoped for but everybody seemed to enjoy it.

In the morning (Sunday 24 June) Bryce, was awakened by Moulaye Ahmed calling in his ear, having reckoned it was time to get up. He was a very nice boy and Bryce gave him and his friend, who had come along with him, each a kite as a present. The owner of the grove came about 6.30 a.m. and served tea, the group collected their things, climbed back into the Land Rover and returned to the gîte for breakfast.

Most people went to bed and rested, but others worked on the problem of plate contamination, in the belief that the prime source was the aridity of the climate and the ease with which static electricity could build up.

They could not do any astronomical work because the day became stormy, but they did find the interesting fact that the whole telescope assembly showed no electrical resistance, and formed a perfect Faraday cage, so that that any electrostatic phenomena must be taking place inside it. After supper, the wind died down somewhat, and they all went to see the skits performed by the local children, originally intended in honor of the Minister of Education, but he did not appear. Moulaye Ahmed was one of the chief actors. After that, Burton and Bryce went to the hut determined to track down the demon, but the roof had only been opened for half an hour when the wind rose, and dust began to blow in. In hopes of doing some astronomical work they slept in the building, getting up twice to check the wind, but by 3.30 a.m. it had not abated and they went back to the gîte.

The wind was still raging next morning, (Monday 25 June), and the sky was full of dust. Dick Mitchell and Richard Matzner were already at work, and the generator was going, supplying power to all the electrical equipment except the air conditioner. They left it running so as to estimate the gas consumption, an essential piece of information for Mikesell at the November return.

Among other things, Gorel spent some time with the Préfet and learned that the episode with the adjutant was the second which had occurred recently. The Préfet was very annoyed and was asking the Governor to have the adjutant transferred to another region. He also learned that the Office of Tourism had reached a decision on the price of hiring camels, namely a thousand francs per half day, two thousand for a whole day, which seemed very reasonable.

At lunch time, a great crowd arrived, beginning with Ambassador Murphy and Ronald LaCount. The latter brought a letter giving them permission to make a movie of their research effort for the BBC, and brought ten rolls of film, which could be used with NCAR's movie camera. The French scientists from Atar also arrived, including Rösch, though they did not meet him as he stayed in his room until the others had left to go to the site. During the afternoon an NBC man photographed the outside of the building, and then a crew from the French department of higher education came, and photographed the inside, and asked Bryce to give a little talk in French explaining the equipment. All this time the wind was blowing furiously outside and sand was sweeping over the ground. When the Ambassador and LaCount came in, while the camera crew were still packing up, Bryce was secretly reminded of the famous Marx Brothers scene in "A Night at the Opera", when dozens of people packed into the same small cabin on a passenger ship. They took the occasion to thank LaCount very warmly for the support given by NSF, both directly, and though NCAR.

As soon as the visitors had gone, they ran through a regular fire-drill, but did not open the hatch because of the wind blowing outside. Then over to the school for the *distribution des prix*.(Figure 75).This was the occasion on which the T-shirts, pencils, pads etc. purchased in Austin with money given by many people, as well as the T-shirts given by NCAR, were given out to the students. The NCAR T-shirts, being fewer in number and of different design, were given to the top students, who had earned first, second, third, fourth, or fifth prizes in each class. A different member of each team was asked to hand out each of these prizes. In addition, Cécile presented some books on astronomy and other items to the school. The ceremony lasted more than an hour, and was quite picturesque, but unhappily the students had to sit on the ground in the blowing sand and wind, and they became somewhat restless before the end.

All the team were so hot and dusty that they repaired to the bar for a beer or Pepsi, where they encountered a young man from the Smithsonian Institution, who had just arrived in search of the fabled meteorite of Chinguetti, first reported some fifty or sixty years before, by a French officer traveling in the region. He had been led to its location on condition that he bring no compass and take no notes, and had reported a meteorite some hundred meters in length and forty meters high, making it by far the largest known, with a mass of the order of a million tons. People had been searching for it for years, and he was the latest. (Welton et al., 2001). He told them that he had previously visited the strange formation at

Richat, and said unequivocally that it was neither volcanic nor meteoritic in origin, originating from a peculiar dome-like structure in the Earth. (Figure 69). He was an interesting individual, but very condescending with a tendency to get on people's nerves.

Later in the day, they learned that the French group, including Rösch, had visited their building about four o'clock, just after they had left for the *distribution des prix* and had been let in by Claude Morel. One reason that they did not see more of the French was that they had arrived unexpectedly, without warning Mr Anezin. No food was prepared for them, and the dining room was already filled with the other visitors. So they were supplied with bread, cheese, wine, and other things to eat at the site.

To their dismay, the team found that one of the Accutrac drives had gone out of commission, and would have to be used only in emergency.

Cécile had spent the afternoon with Sidi Ould Habott going over the 17 th century astronomical manuscript that she had photographed, and had persuaded him to a joint effort of a report on it for some learned journal. (This proposal never apparently bore fruit). By this means Cécile hoped to get access to the library catalogue. Sidi Ould Habott was the heir to the library, which had been catalogued, but the Habott family never permitted outsiders to copy it.

After supper she went with the camel boy, because three of the group, Chuck, Richard Matzner, and she herself, were due to go on a camel trip the next day. At the first night shift, Burton and Bryce were to work on tracking the electrostatic demon, while Gorel and Mitchell were to do limiting magnitude photography on the second shift. Finally Chuck and Richard Matzner were to come in at dawn to calibrate the photometer, and then leave immediately. Unfortunately Gorel became ill during the afternoon with a high fever, with extreme intestinal upset. He was treated by Dr Gillette.

When Burton and Bryce went to begin their observations,

the wind had died down somewhat, but was still strong enough to bring dust in when the hatch was open. So they opened it only very briefly for each exposure, and kept the lens completely covered with the thermal blanket except during the actual exposure period. This turned out to have a profound effect on the instrument. When the first plate was developed it was found to be completely out of focus, a terrible discouragement, because the true focus had been found several days before, and was consistent from night to night. There was a new filter in position to replace the old scratched one, but this seemed hardly likely to cause such a change, and they were going to check that anyway. This was the first time without the lens having been exposed to the night sky for a considerable length of time. Perhaps on previous nights the lens temperature might have been significantly different from what they thought it was, since they had only recorded the temperature inside the building. This time they recorded both the building temperature and the thermistor readings which Mitchell could convert into temperatures.

Examination of the first plate led to the discovery of what they thought was their electrostatic demon. They had put moist sponges in the telescope tube to increase the humidity, but on the first plate they could see the sharp edge of the plateholder frame and a kind of diffuse blackening extending outwards from this edge with no apparent relation to the sector operation. Perhaps moisture from the sponges was controlling that, if it had ever existed. The outline on the plate was precisely that of the plateholder frame and not that of the back end of the telescope, and did not include any trace of the little metal baffles added to cover the step-wedge portions of the emulsion. So, could the outline have been produced during the process of loading the plate, and not during the exposure? This question was quickly answered, because another plateholder had been loaded earlier in exactly the same manner by Burton. The unexposed plate was quickly developed and showed exactly the same phenomenon, so that the trouble lay in the plate loading process, possibly from the use of the Static Master brush. They loaded a plate without using the brush for a focus run. It came out perfectly, so henceforth the brush was not to be used, but the telescope was so far out of focus that they would have to take another plate. This they could not do that night because there was too much dust.

At 5 o'clock on the morning of Tuesday June 26, Cécile left to join Chuck and Richard to begin their camel ride. The camel boy had arrived with three camels at 3.30 a.m., and been asked to wait at the observatory until it was time to depart. The camels looked quite handsome, with fine saddles, well provided with padding to ensure comfort on the long ride. The plan was to go some ten kilometers up the line of oases northeast of Chinguetti, along the edge of the sea of sand. (Figure 76). The others slept and puttered about doing odd jobs.

Cécile, Chuck and Richard returned between five and six o'clock after a long camel ride. They had not stopped at the palm groves, but, at the insistence of the camel driver had gone some 16 or 18 kilometers into the bush to a place where there was a well, some thorn trees for shade, and forage for the camels. It took four hours each way, and as Chuck Cobb recalled in his account already quoted, had included a *méchouie* of goat meat. They would have preferred to stop in the palm groves, but the people who own them didn't want camels in there where there is no forage for them. Also, no special arrangements had been made to receive visitors at one of the groves and the rules of desert hospitality required that the owner or his representative be ready to provide tea or even a meal.

Bryce described the evening's work by himself and Burton as the most humbling experience of their whole stay, at last learning some of the things they had to look out for. The first plate for a focus run, showed the ghosts even though the polonium Static Master brush had not been used. Moreover

Figure 76: Cécile aboard a camel.

Figure 77: A typical desert herd of goats.

the focus was nowhere near where they thought it ought to be from the experience of the previous night. They found that the focus of the lens changed by as much as two millimeters as it cooled down on exposure to the night air. This effect did not show up on the thermistors because they did not record the cooling effect as rapidly as it was taking place, not being in close contact with the critical components of the lens, but it was too late to do anything about that. The prime cause was the change in temperature of the framework holding the lens components in position.

They had to do yet another focal run, this one admittedly temporary, because there was no time to heat the building up again and wait for equilibrium, and they also very much wanted some limiting magnitude plates on various emulsions. The III-O's turned out fairly well, though showing a few spots, cause unknown, but the IIa-O plate was a disaster from the point of view of quality, though adequate for limiting magnitude determination. This time the ghosts reappeared with a vengeance, from evidence, thought to be due to light leaks on Johnny Floyd's plateholders. They had seemed so well designed that they had not been extremely careful to shield them from light, but part of the problem could be because Chuck Cobb had shimmed them a little to make the dark slides draw more easily.

Because of all these problems they did not finish work until well after 3.00 a.m., and then went back to the gîte to fetch Richard Matzner to try to contact Al Mikesell, but after two hours he had not succeeded. Gorel and Mitchell came at about 4 o'clock to begin calibration field exposures, postponed until the new filter was in place. For this they needed a focal setting that worked, and the best that could be done was to give them the one determined earlier in the evening. However, with so many people in the building;—Bryce hoping to talk to Mikesell, Richard Matzner operating the radio, Burton developing plates, Gorel and Mitchel changing battery voltages to the sensitimeter box and getting set up for their calibration-field plates—things did not go efficiently, and dawn arrived before Dick and Gorel could make their exposures. With all the difficulties and anxieties of the night, Burton and Bryce did not get to bed until a little after 5 a.m., but were up again at 6 to start their camel ride. Dick Mitchell did not want to go, and Gorel was not particularly interested, having made many long camel journeys when he was younger, so Cécile went with them instead. The camel drivers and the little boy were the same who had taken Chuck and Richard into the desert, but this was a new day, (Wednesday June 27), and a new destination.

Though their minds were full of worries about the scientific situation it was a pleasure to be out in the wilderness. They set off in the cool of the morning, and the scenery was often spectacular. Once they left the village behind, they got into the real dunes with their ever-changing shapes, with an occasional flock of goats, (Figure 77), and a few sheep. For a long time they followed the bed of a wadi, finally halting at a group of wells, with a few thorn trees, which, however, provided good shade. Burton and Cécile settled under the biggest one, while the camel driver prepared tea. Bryce went under a smaller one, where he tried to sleep, which they all did a little, though bothered by flies.

The weather, from the eclipse standpoint, deteriorated during the day with a moderate but strengthening wind, clouds in the sky, and a lot of dust, churned up by distant great thunderstorms. Bryce felt very pessimistic about the weather prospects for eclipse day, for they had never yet seen in Chinguetti a sky as good as that of the brief visit a year before.

The return trip was fun, with the riders guiding their camels themselves. This was done with a stick and a rope attached to a ring in the camel's nose. With the stick, the rider tapped the side of the canel's neck opposite to the one to which he wanted the animal to turn. For increased speed he tapped the neck with his feet and shouted "ayt,ayt, ayt", which, if done strenuously enough, put the animal into a trot. The walking gait was quite different from that of a horse, but the trot was rather similar. The trip was shorter and less tiring than the one taken by the others the day before, only perhaps some eleven kilometres into the back country, with no *méchouie*, but a few provisions brought from the gîte, which they shared with the drivers. They shared their water, which had been drawn from the well where they stopped and put into a goatskin guerba. It was a lot cooler than their Evian water.

At the gîte at about 6.30, there was a message from Mikesell on the subject of electrostatic problems, and some mail from Nouakchott for Cécile concerning the bank account she had opened for expedition use with funds from her French bank. The Nouakchott account helped other team members, and even, on occasion Medrud himself, when they wished to change money. Finally there were some people from the Mauritanian radio network, who wanted Cécile to act as interpreter in a request they wished to put to the leader of the American expedition, namely Medrud. Cécile declined, pointing out that Claude Morel was the interpreter for the NCAR group. What they wanted was to be able to use some of NCAR's power on eclipse day, so as to be able to transmit a 30-watt broadcast. This was not much power, but Medrud was reluctant to let them on the line until a trial run saw how things would go. There was to be a big dress rehearsal the following morning with everybody going through the steps scheduled for eclipse time, to see whether NCAR would have any problems in delivering the necessary power to the various groups. The leg of the three-phase circuit that Texas was on, was the one drawing the least power, and Medrud had asked them to leave the battery charger and refrigerator on during the eclipse. The air-conditioner, the cooler, and most of the lights were to be turned off and they would be running mainly on batteries, with only the fast drive receiving NCAR power.

Unfortunately the Mauritanian people would not be ready to set up their equipment by the time of the rehearsal, and Medrud was reluctant to give them approval. Bryce said that they could use the Texas portable generator. Another request came from the German group in Atar who asked for some photographic plates, their own having been stolen.

Those had an infrared sensitive emulsion of a special type, but rather than folding up their camp and going home, they felt they should go through as much of their experiment as possible using whatever plates they could get hold of. The Texans had more than they needed of 8 x 10 inch IIa-O plates, and would give the Germans some. They didn't need such large plates, but would take three or four, and cut them in half. The message was received by Medrud and passed on, with no explanation, to all the American teams. The Germans would come in the next day when the Texans could talk to them.

While the others were out in the bush, Matzner had used the NCAR movie camera and some of the film brought by LaCount to do some expedition and village photography. He would film the dress rehearsal, and the tasks of the various team members, on the next day, but on eclipse day itself nobody but the team would be allowed in the building. At supper time Chuck Cobb suddenly felt quite ill and had to go to his room, accompanied by Dr. Gillette, who thought that his feeling faint had been brought on by the long camel ride, and exposure to heat of the day before. This brought up another worry for Bryce, who prayed that all the team would be fit on eclipse day.

He had originally been scheduled for the evening duty with Burton, to continue focus runs, this time holding the temperature of the tube and lens exactly at 85 °F, the temperature expected at eclipse time. He felt extremely fatigued and Cécile went in his place. Burton was also tired, but he seemed to have an iron constitution. Bryce was most grateful for his presence. The weather did not look good, there were still clouds but some stars were visible. Perhaps some work could be done, but this was cutting things awfully closely, and if conditions did not improve, perhaps they would have to omit the calibration plates. If they were not done at eclipse time, perhaps they could be got the next night, which would entail making a move of the telescope, as slight as possible, from its eclipse position. The entry for Thursday June 28, records that quite a bit was accomplished in spite of adverse conditions. It was overcast during the first part of the night, but while Burton slept, Cécile put her head out every half hour to look at the sky. At 2.30 a.m. it cleared off and she at once went and called Richard Matzner and the rest, who were scheduled for some sort of work during the night. Focus runs were the first order of business, and showed clearly that the telescope was stable if the temperature was kept constant. From then on the temperature would be 85°F, estimated to be at, or slightly below the ambient temperature for 10.30 a.m. How Al Mikesell could duplicate this in November when the days would be much cooler, Bryce did not know, but that would be the newcomers' problem, not his. Not only did they get good focus plates, but they also aligned the five-inch guider to point at epsilon Geminorum during the eclipse. Mitchell, Matzner and Gorel also got good calibration plates before dawn. Bryce intended to go through the same routine the following night, and then leave the telescope untouched until eclipse time.

Fatigued from the stresses of the previous 49 hours, Bryce slept all night, but awoke with a slight cold. Between 10 and 11 a.m. they had the full scale rehearsal, with all the teams participating to see what power requirements during the eclipse would be. Everything went off perfectly, with Richard Matzner shooting some movie footage, with the roof hatch open to let the Sun in, to give enough light for the Kodachrome-II film. He also took a shot of Cécile on the roof acting as shutter man. During her night vigil she had made a brand new cover of several layers of aluminium foil to cover the lens end of the tube, and she pretended to remove it, though she could not do this in reality with the Sun shining directly down the tube. Richard also took a shot of the hoisting of the McDonald Observatory flag.

At lunch-time there was a general discussion of all the things which had to be done at the eclipse, and Bryce made final assignments. The place had filled with visitors, including Mauritanian radio, the CNRS, (the French national research organisation), French radio and television, and the French Department of Higher Education.

In the afternoon most took a siesta, but at 3.00 p.m. Cécile went with Richard Matzner to pick up Sidi Ould Habott and another school teacher to shoot some movie film including a record of Cécile photographing books. The rest went through a fire-drill with Chuck Cobb calling the orders, and Bryce taking his place at plate-changing. When Cécile and Richard returned there was another drill with Cobb giving the orders and Bryce taking Matzner's place reading the photometer, with him joining Burton at plate-changing. This drill was to make doubly sure that Gorel would understand correctly and act quickly on the commands that Burton gave him during the ten minutes before second contact, when the two of them would be trying to line up on the guide star, of uncertain visibility on the bright sky. Everybody felt that they had been through the routine so many times that further practice was unnecessary, and no help to relieving the tension of the eclipse moment.

Afterwards, Chuck replaced the deionizer with a new one, the first being exhausted, the new one to be used for the eclipse plate water. It would not be used for any other purpose until November. The fifty cases of Evian water had arrived, and since then they had been doing the photography with untreated water. For the eclipse plates it would be run through the deionizer, with a completely separate set of chemicals for each plate, which meant running 96 bottles, or 8 cases through the deionizer.

While this was going on, Bryce occupied himself with clean-up duties and helped Richard Matzner attach a grounding wire to the telescope tube.

They did not, in fact, think there would be any real effectiveness to this, but Mikesell had suggested it.

The plates taken the night before turned out very well, extreme precautions having been taken in every step of the loading process, including inserting wet sponges in strategic places to counteract any electrostatic discharge. In this regard Bryce noted that the weather had become more humid, with a wind from the south and distant thunderstorms which had lifted a vast quantity of dust into the sky. The sky looked exactly as it did when he was in Niger the previous year, typical of Agadès, which he had rejected at the time. He was sorry that the prediction for Chinguetti as a good site had turned out in the end to be false. In all the weeks they had been there, they had never once had as good a sky as that seen there the previous year. They had had some usable skies, they had not always been bad, but he feared that the Intertropical Convergence had come so far north that, failing rain, they would have dust.

Just before supper, Cécile went to see the Préfet to discuss the problem of a guard for the building over the long wait. The following arrangement emerged :—The present night guard, a very nice man, would remain as a permanent 24-hour guard. Originally, the Préfet had thought in terms of an armed guard, at a time when he thought that the expedition had no money to pay for the service, but since the NSF was willing to pay, he preferred someone from the village who was in retirement. The guard was to move his whole family, his tent and his belongings close to the building so that it would be watched at all times.

She returned after supper accompanied by the Lillers, astronomical visitors from California, to discuss the disposition of some presents for young children that they had brought. At the same time, Bryce was at a meeting of team leaders called by Medrud, to discuss how long the reserve emergency power should be kept running. At eclipse time, all the required power would be supplied by one of the diesel generators, while the other would be running free with no load from 10.15 to 11.30 a.m., ready to be thrown on line if anything went wrong with the first generator. Medrud wanted to keep the period of running on no-load down to a minimum since this would be very hard on the machine.

Thereafter, Bryce went with Burton to do a final focus run, the building having been kept at 85°F for a long time. The results showed the focus in perfect adjustment. Then came final alignment of the guide telescope. It was fortunate that they were able to do all this in spite of dusty skies, though, as in Agadès, the dust hung low and stars at a higher elevation came through well, with many on the focus plates.

Richard Matzner, Dick Mitchell and Gorel went out in the early morning hours after Bryce and Burton had quit for the night, to try to expose some calibration-field plates, but the wind got up, throwing a lot of dusty material up in the sky, along with clouds, making it impossible to see any stars. Only one calibration plate had been obtained, with too few star images to be of much use. Even so, Bryce decided not to try for one after the eclipse, but to leave the telescope untouched until Mikesell arrived.

Friday June 29, Bryce got up about 9.00 a.m. and found Chuck Cobb and Burton Jones already at work getting the darkroom in order and preparing chemicals for processing the eclipse plates. Chuck was already passing the eight cases of Evian water reserved for the processing work, with a ninth in reserve, through the deionizer, refilling the empty bottles with the treated water. These were being stored in the darkroom at standard temperature. The air-conditioner had been very good at keeping the darkroom temperature at 68° F, and they could only hope that it would keep doing so through eclipse day. Meanwhile the rest of the building was being kept at 85°F by running the cooler intermittently, and every so often turning on a hot plate borrowed from NCAR.

Bryce spent the morning vacuuming the building, straightening things up and helping Burton and Dick Mitchell who had arrived, to align the Questar. They mounted it on the cradle of the main telescope and then put a very small pinhole over the big lens. At about 10.30 a.m. the Sun was in a good position through the roof and with the big telescope centered on the Sun and a developed plate mounted in back, centered the Questar on the Sun. (See figure 56). The Questar, now with its Sun filter in place would be used as an emergency guide telescope if they failed to pick up epsilon Geminorum on the five-inch guider ten minutes before eclipse. In that case they would aim directly at the crescent Sun. The front and back ends of the telescope were by now covered with thermal blankets made from several layers of aluminium foil.

During the morning, Cécile went to the Préfet with Medrud to make final arrangements for the guard to watch the building during the long wait. He had typed up two copies of a statement that Cécile had prepared, one to be signed by him, the other by Medrud, detailing the arrangement and specifying the amount to be paid to the guard.

While she was there, she found out that Medrud had not given them some important information. Last night, while with the Lillers, Cécile had told the Préfet that Medrud had been emhasizing to everybody that nothing that was not originally specified on his customs manifest as remaining behind in Chinguetti could be left, and that the customs people were very strict. The Préfet replied, however, that Bill Curtis had told him that any equipment or material that could be left in Chinguetti, would be left. The Préfet and Cécile were anxious to have this matter straightened out before they all left, but the Préfet did not know how to bring it up with Medrud. They agreed on the previous night to try to bring the subject up in conversation in the morning. Because no suitable moment had been found, Cécile simply turned to the Préfet, and asked, "Is there anything else you would like to discuss before we leave?". After a momentary hesitation he said "Please tell Medrud that if he has any instructions to give me about things remaining behind, I should be glad to carry them out". Cécile translated this to Medrud, but there was no response and tried again, to which Medrud replied, that, in fact, a customs man would be sent to Chinguetti to make arrangements for the things that could be left, and that, if the man could not come, appropriate instructions would be sent to the Préfet. The latter replied that there would be no problem since he already had authority to act as a customs agent.

On the way back to the gîte with Medrud, Cécile thanked him heartily for this new development, namely that because of the presence of an official customs agent it would not be necessary to adhere strictly to the original manifests, and there would be much greater flexibility. She asked him how this came about, to which Medrud answered that he had simply got the communication by radio a few days before. It was a pity he had not told Cécile earlier, because the previous night she had seemed rather stupid in front of the Préfet, who had already been told by Curtis the same thing that Medrud was now finally telling her.

After this Cécile went to the infirmary with one of the teachers who was bringing his mother along. The mother was ill, but the (male) nurse at the infirmary had previously refused to allow her to see the doctor or receive treatment. Cécile went to see whether she could prevail upon the nurse to let Dr. Gilette see the woman. The nurse made a scene complaining that they were trying to bypass his authority, but eventually she was seen by Dr. Gilette and found to be seriously ill with high blood pressure. This was worthwhile even though he could not give her the treatment which she would have got in a real clinic. He could only advise her on diet, and tell her that if she became worse she must go somewhere where she could get regular medical attention. Everyone left in a good humor, even though Dr. Gilette had refused a previous request to see the woman outside the regular clinic hours, because the nurse, who ran the infirmary, had expressed annoyance at being bypassed when he had seen other people outside regular times.

At lunch, the team all signed their two remaining photographs, provided by Harlan Smith, of the Earth seen from the Moon. At Gorel's suggestion one was given to Mr. Anezin, the innkeeper of the gîte as a mark of gratitude for all the many things he had done for them. The other was to go to the Préfet.

During the afternoon, Chuck and Burton continued with preparation of the chemicals for the next day, while Richard went about taking more snapshots and shooting movie film. Bryce finished cleaning up the building, verified that all telescope components were GO, checked the batteries, and generally took a look at all the systems in the building. Gorel thoroughly wiped the telescope, removing all traces of dust, being most careful not to touch the guide telescope or the Questar.

Cécile spent part of the afternoon packing, and part with the school teacher who was making the translation of the manuscript from the Chinguetti library. Bryce left the building and wandered round the village, going first to the tent belonging to the family of one of the men who had been on the camel ride the day before, where he expected to find Cécile having tea. He did not find her but had tea himself. Mohammed Ould Dick, the little boy who had accompanied them on the camel trip arrived, at first unrecognized by Bryce, because he had had his head completely shaved.

After tea Bryce went with him to a palm grove nearby which belonged to his family. This was abandoned and in a state of decay, because the black servants who normally tended it had run away. Nobody was drawing water from the well, the trees were not being irrigated and were dying. Then they went to the tent of his aunt, who, he discovered was one of the schoolgirls that Cécile had dined with weeks before, whom Saidu had brought into their building one night. They had since learned that she was one of the village prostitutes. (Some of the village elders had warned Cécile against further association with her). She made eyes at Bryce, and he thought that Mohammed Ould Dick was trying to set up a tryst for the two of them, but he pretended not to understand and played instead with her 18-month-old baby. From thence to Mohammed's house, where Bryce tried to fix up a kite that he had given him, which had broken.

Then to the gîte where he met Moulaye Ahmed Ould Sidi Ali who was sitting by the gate. He invited Bryce to come to his house, where, in the courtyard, he proudly displayed a model, about three feet wide and one and a half feet high, of a nomad tent that he had made. He had shaped wooden sticks in the form of tent poles and sewn together pieces of cloth for the fabric. Bryce was extremely touched at the gift, and felt sad that they would be leaving Chinguetti in less than two days, without having had more time with the people.

At supper, everybody was quite relaxed. They had done everything they could and all was ready for the morrow. The Intertropical Convergence seemed to have arrived for good, and there was dust everywhere in the sky, so that chances of success seemed very remote.

Dick Mitchell went on the evening watch to maintain the temperature at a constant 85°F. At 2.00 a.m. Richard and Burton would take his place to prepare the plate sensitometry images and load the plateholders.

## E-Day and After

Bryce got up at 5.00 a.m. and went to the building, meeting Burton returning to the gîte. The weather looked unpromising with overcast in the west, and though the clouds moved off a little later, the dust was heavy in the air on all sides. His first job was to go to the NCAR radio shack to set his chronometer from the WWV time transmission. He then went back and checked the battery voltages and all the systems in the building. The only one at fault was the spare Accutrac which was not putting out the correct voltage. He was reluctant to switch it on early, although in principle it should be done at least two or three hours in advance to achieve stability. Burton and Matzner had finished the pre-exposures of the plates on the réseau and step wedges, and they were ready in plateholders wrapped in black cloth.

At about 8.00 a.m., Cécile arrived at the building and Bryce went to breakfast where everyone finally gathered. At about 9 a.m. they were all at the building, bringing some exposed and developed photographic film through which they could look at the Sun. At 9.30 a.m. they could see the beginning of the eclipse. Bryce had put some film over his binocular lenses and got an excellent view. Although the eclipse was clearly visible, the astronomical outlook was discouraging. High thin clouds covered the Sun; a great sandstorm was blowing, and it was very hot. Nevertheless they were determined to go through the motions of their familiar eclipse routine, even though opening the roof in such conditions was likely to bring large quantities of sand into the building.

He was worried about the possibility of Cécile's losing control of the shutter while on the roof. But the eclipse, or their part of it went off without a hitch. A full report of that part of the day is on a magnetic tape which recorded the sounds in the building, before, during, and immediately after the eclipse, as well as comments made at the debriefing session after lunch. The most noteworthy thing to report was that the wind dropped dramatically about ten minutes before second contact, so that when the roof was opened, very little sand got into the building, and Cécile on the roof had no difficulty with the shutter.

The sky was so bright with its great load of dust, that it was impossible to find the guide star even up to thirty seconds before contact, and the emergency routine using the Questar was followed. The sky was brighter than it was during the three-minute total eclipse that Bryce had seen in North Carolina in 1970.

As soon as the telescope was moved back into eclipse position, after the initial comparison-field exposure, Burton could see epsilon Geminorum in the five-inch guider, close to the cross hairs. They were a little off, perhaps 4 mm in one direction, 1 mm in the other, but they still thought they would get two of the three stars they wanted.

Venus was clearly visible to the naked eye during the eclipse, but few other stars were, though the fact that Burton could so easily see epsilon Geminorum in the five-inch gave hope that they might have a fair number, perhaps a dozen star images, on each plate. This would be enough to justify a return to Chinguetti in the fall. Another good sign was that the solar corona was not very extended, which augured well for the chance of recording stars near the Sun. (See Figures 3 and 6).

Before returning to the gîte, Gorel and Bryce went to see the Mauritanian radio people, who were transmitting live from the site to their network. Bryce gave a little account of the Texas project, and noted their activities in Chinguetti, both at the library and at the school, and took the occasion to thank the Mauritanian government and the people of the oasis. While he was doing this, Cécile was off, taking her supply of eye ointment to the tent where the little girl with the badly infected eye lived, leaving it there with instructions for its use.

Most of the other teams were quite happy with their results in spite of wind and sand. The Texans held their own debriefing session at 11.45 a.m. and then had a leisurely lunch. After that, Chuck Cobb, Cécile and Bryce got together to make a list of things to be left in Chinguetti, if the experiment turned out to be a failure, about which they would have to wait until the plates had been processed and thoroughly dried, for examination. The list would only be tentative, subject to Harlan Smith's approval, if the experiment was a success and everything was left until November. Bryce had asked Frank Oral to come to his room, to let him know that after Bryce went, Chuck would be in charge, and that they were leaving a large amount of equipment behind, following the news of the liberalization of the customs agreement conveyed to Cécile by Medrud the day before. He hoped that the news about the customs situation could be conveyed to the other team leaders. Frank indicated to Cécile that he would soon tell her how much his team could contribute to the cost of the gifts given to the schoolchildren.

The DeWitts then went to the observatory to pick up a few things, together with some cans of paint and a paint brush for Moulaye Ahmed Ould Sidi Ali in return for the model tent he had given Bryce. Sidi was planning to paint a new room, of stone and banco, just built onto his house, and he might be happy to get the paint which would not be of much use by November after sitting around for five months.

They met the camel boy, Mohammed Ould Dick, who invited them for tea, an invitation that they had to decline as they were busy packing. Bryce felt a rush of sadness at telling him that they would be leaving in the morning, and that they had not had more time for the people of Chinguetti. Mohammed had often followed them around, not only into the desert, but in their comings and goings in the village. He had enormous style, not only in appearance in his long flowing boubou, but in his manner of speaking. He loved to recite long passages of melodramatic French poetry by such as Victor Hugo and Baudelaire. At their news, the boy turned away to hide his tears. At parting the DeWitts were each given a small gift. The rest of the afternoon was spent packing, with interruptions, by visits from the school-teacher who was helping Cécile translate the scientific manuscript, as well as from the nice driver, the aristocrat from Boutilimit, whose address Cécile wanted to get if they should ever return to Mauritania.

After packing, Bryce went to look for Moulaye Ahmed Ould Sidi Ali, who was supposed to be at the gîte about 6.30 p.m. He found him and Mohammed Ould Dick at the gate and gave him the paint and paint brush and a roller, showing him how to pour the paint in a pan. He returned to find something for Mohammed and gave him the pith helmet he had worn for 27 years, the shoes he had used in Chinguetti (for the leather), a whetstone, some remaining Chinguetti buttons, and a pencil. Mohammed gave him some arrow-heads and some to pass on to Burton.

Just before supper, Burton and Richard Matzner came in from developing the plates. Each had been processed in a separate chemical solution and they reported that they looked promising, with some star images. Bryce told Burton about the gift of arrow-heads and went to fetch them, and then he went with Cécile to thank Moulaye Ahmed.

At supper, they ordered two bottles of Moët et Chandon Brut Impérial from Mr. Anezin for a happy gathering, everyone feeling that their efforts had been worthwhile. Most of the other teams knew that they had substantial results, whereas the Texans would not know for weeks or months to come, but were expecting to have some idea later in the evening on examining the plates. Even in spite of the poor sky conditions, Burton had reported good seeing through the guide telescope, so that if there were any star images, they might be good.

By evening, the wind, which had come up again strongly after the eclipse, had died down, though the air was still full of dust. It had been an exceedingly hot day, perhaps the hottest they had ever had, with 112 ° F in the shaded and protected enclosure where they kept the hygrometer. This meant that the temperature at which they had carefully maintained the telescope during the previous 24 hours, namely 85 ° F, was some 12 ° F lower than the lowest temperature reached by the air near the ground during the eclipse. Before the eclipse the temperature was something above 100°F and fell only to 97 °F by mid-totality. Bryce had noted as miraculous that the wind suddenly dropped during the eclipse, but he now thought that a shadow 250 miles across, coming from the west might have had a profound effect on wind blowing from the NNE.

Bryce records in his diary the role of each member of the team during the eclipse. Cécile was on the roof acting as shutter man: Dick Mitchell and Chuck Cobb opened the hatch and helped Cécile hold the shutter against the wind until she was in a secure position. Chuck Cobb and Burton Jones acted as plate-changers, one of them removing the plates, the other inserting them. Burton, also had the job, together with Gorel, of aligning the telescope in the moments just prior to the eclipse. Dick Mitchell kept a record of the indoor and outdoor temperatures and the temperatures of various parts of the tube, obtained from the thermistors. Richard Matzner measured and recorded the sky brightness with the aid of the photometer. It was he who gave the command to use nothing but the III-0 plates, of which altogether three were exposed. Had the sky been darker, they would have exposed a III-O plate at the beginning of the eclipse, another at the end and two IIa-O plates in the middle. Gorel had the crucial and delicate job of changing the declination, operating both the clamp lever and the declination shift lever, after adjusting the declination-stop set-screw during the initial aiming of the telescope. Bryce's task was simply that of calling out elapsed time and giving the various commands at the appropriate moments. The whole thing went off more smoothly than any of the fire-drills had ever gone. Everyone had set his heart on doing the best possible job, and everyone performed splendidly. This was an expedition in which everyone did his very best. They could not have done better.(Figure 78).

Burton woke Bryce at midnight and they went with Richard Matzner to look at the plates. They looked quite well considering the condition of the sky, and a quick check against Dick Mitchell's star charts revealed at least a dozen eclipse-field stars, maybe twice as many on each plate. This was enough to make them potentially the most valuable eclipse plates ever. The decision was thus made to come back in November for the second-epoch photography. The plates were put in a special box constructed by Cécile, who would carry them with her at all times.

On leaving the building around 2.00 a.m., they said a final good-bye to the night guard, who knew that three of them were leaving in the morning. As they shook hands, he gave

Figure 78: Globetrotters triumphant!
Back row: Bryce and Burton Jones.
Middle: Alassane Sy, Matzner and Mitchell.
Front row: Cécile and Chuck Cobb.

Figure 79: The hut is handed to the care of the guard.

Bryce four little bones, vertebrae of a goat that his daughter had given him for Cécile, who had been collecting them because, as a girl, she used to play the French version of jacks with them. As his hand closed, the guard said "ouada'na k'moulana", which is "Farewell."

Four members of the team left on July 1, taking the eclipse plates with them, leaving three to secure the building, ready for the follow-up team. They worked the rest of Sunday, Monday and Tuesday, putting as much equipment as possible inside, removing the hardware from the doors, putting plywood panels over them, and caulking and nailing them in place. The hut was now left to the care of Bah Ould Soueidi.(Figure 79).

On Tuesday, the leader of the NCAR support team said that he had doubts about the aircraft being able to take off with the expected load from the short runway at the local airport, and called for volunteers to ride in a Land Rover to Nouakchott. Cobb was the first of many volunteers, and eventually six rode, leaving their baggage to come by plane. They were scheduled to leave at 6.00 a.m. on Wednesday July 4, actually leaving four hours later, and, as Cobb puts it, "Drove through some of the most rugged country I ever hope to see. We soon left the sand dunes behind and there was nothing but jumbled rocks for miles. The mountains were even more barren than in the Big Bend country of Texas. The road was in some places little more than a trail with rocks piled in the dips so a Land Rover could pass. We met some cars, but were glad we were in a four-wheel drive vehicle."(Figures 80 & 81).

They passed through the small village of Atar, and then Akjoujt, a small mining community of 700 or 800 people, after which the road improved and they reached Nouakchott, taking 11 hours to cover 250 miles. This was all new to them, because they had come in by air. After dinner in a European-style hotel, they went to the home of a U.N. economic advisor assigned to Mauritania. They spent about three hours talking about Mauritania and its economic problems. He had only

Figure 80: Descending the Amogjar pass.

Figure 81. Chuck Cobb was glad of a robust
vehicle in such rough country.

been there a month so was not too well informed, but they learned more from him about the country than from anybody else they had been able to talk to. After a few hours' sleep and breakfast they visited the American Embassy to discuss the problem of money. They had heard that the customs people were confiscating all African money. Mauritania had withdrawn from the African Franc Zone on June 30, the day of the eclipse, and had instituted its own currency, the ougiya. One could hardly believe that this date had been chosen to cause the maximum inconvenience to thousands of foreign visitors, more likely that, some time previously, some fiscal authority perhaps quite uninformed about the eclipse and its impact, had chosen this date as the convenient mid-point of a calendar year. The Embassy had not been informed of any confiscation, and assured them that they could take out the money that they had, and that Embassy people would be at the airport before take-off to help in any way they could. The customs people did, in fact, search the baggage and asked a few expedition members what kind of money they were carrying and if they had any African francs. A man from the Embassy assured the customs agent that they were Americans carrying dollars, so the rest of the group were passed through with no questions, to board the DC-3 flight to Dakar.

There they were bussed to the hotel Su-Nu-Gal, with its attractive swimming pool. Although it had rained in Dakar, there was still a water shortage, with the water turned off between 9.00 a.m. and 6.00 p.m., so they spent the rest of that day in the pool. Next day Matzner and Cobb caught a ride to the Artesan Village between the hotel and downtown Dakar. This was a cooperative enterprise of many booths selling the goods and crafts of Senegal with masks and wooden figures, jewelry in silver and gold, leather work, basket weaving, clothing and pottery. Behind many of the booths craftsmen were at work producing these goods.

The days in Dakar were spent in excursions to various places of interest. Sunday took them by a short ferry ride to the notorious island of Goree, the departure point of countless slave ships headed for the Americas, accompanied by many scenes of heartless cruelty. After a picnic lunch, they had a swim, cut short when some of the boys caught a baby shark and began playing with it.

They were told that on Sunday all the down-town stores would be closed so they spent the day round the pool. On Monday they took a public bus, visiting down-town shops and seeing the variety of merchandise on sale. On Tuesday they wanted to go down-town again and found an hotel bus to take them. After lunch with some other hotel guests, Cobb set off to go back to the Artesan Village. Failing to find a bus he was walking there when he found his room-mate, Professor Hagen from Penn State University wandering around. He said he was looking for some wood-carvings. Cobb had seen some, but could not remember where, but that there were all sorts of carvings in the village. Hagen didn't want to walk, so they took a cab, and after a session there, eventually found buses back to the hotel.

On Wednesday, July 11, they ate breakfast, put their bags outside and got ready to go to the airport, leaving the hotel at 10.00 a.m. The trip across the Atlantic in the charter aircraft was uneventful, landing at JFK, and, within the hour, Cobb and Mitchell were on a jet bound for Austin, but at Washington the pilot announced that he was having difficulty with his radar and the weather was not good, so back they went to JFK for repairs. Restarted, with an intermediate landing at Washington, Dulles, they arrived at Austin at 10.30 p.m.

The DeWitts had come to Mauritania long before the charter flight, had regular tickets, and with no need to await its return, Bryce had gone off to France and Cécile had already arrived at Austin with the precious plates, where a ceremonial inspection for the benefit of photographers was duly held.(Figure 82).

Figure 82: Back in Austin, examining the plates.
l. to r. Brune, Mikesell, Thompson, Cécile DeWitt,
Evans, Jefferys, Tull.
(The shadow of the arm carrying the rotating sector in
front of the eclipsed solar image is clearly visible).
(University of Texas photo.)

# The Arabic Manuscripts

The two manuscripts photographed by the DeWitts at the instigation of the local scholar, Mokhtar Ould Hamidou, were identified as:-

A theological work composed by Ebi Hilal el Askeris and written in his hand in year 480 of the Hegira (1087-1088)

A scientific work, El Moughalaâ, written by Ebi Maghroâ in the year 1040 of the Hegira, (1670, copy dated 1103 of the Hegira)

Prints of these two works were made and copies distributed to some three or four interested parties. The context suggests that Mr. Hamidou was well acquainted with the contents of the library, and had even participated in a listing of them. It should, however, be remarked that the impression created by the library, in common with a surprising number of others in great houses and institutions throughout the world, is of the custody of revered objects, too revered in fact for their owners to allow them to be ordinarily read. This explains why he was so interested in having the detailed texts before him, though at the present writing nothing is known to these authors that anything ever came of this.

From the astronomical point of view by far the more interesting text is the scientific one. We need to hark back to the history of the access of the nations of Western Europe to the ancient astronomical and other scientific works originating in Greece and even farther East. One route was from Constantinople during its heyday after the sack of Rome by the Visigoths. This continued as Rome was re-established as the center of the Christian faith, continuing until the capture of Constantinople by the Ottoman Turks in 1453. The other lay through the lands of Arabia. Works such as the great compendium of Claudios Ptolemaios in the first Century AD., had been based on observations made with the unaided eye on graduated quadrants in Alexandria, and originally written in Greek. With the rise of Islam in the 7th Century it came to dominate northern Africa and the whole of Spain, being only turned back by crucial battles, such as that won by Charles Martel at Tours in 732, and even later, in the 16th Century at Vienna, commemorated by that bakery delicacy the croissant. Ptolemy's works were translated into Arabic, and formed part of a large corpus of scientific and other learned texts which diffused into the Moorish territories in Spain. The Moorish rule, lasting more than half a millennium was often very tolerant and enlightened, and such sages as Ibn Sina, known in the West as Avicenna translated an enormous body of medical, philosophical, and other learned works into Latin. After Leon and Castile gained a measure of independence, the monarch, Alfonso X, 'El Sabio', the wise one, is particularly credited with fostering the transmission.

By the time of Copernicus, (1473-1543), the Church was so well established in Rome that it could deem his work, which undoubtedly fostered a heliocentric picture of the planetary system, sufficiently at variance with the Ptolemaic picture, now received as doctrine, to foist on it a preface by another hand. Johannes Kepler, (1571-1630), undoubtedly favored the heliocentric system, while Tycho Brahe, a master of naked eye observations, managed the remarkable feat of creating a composite system, part heliocentric and part geocentric. Finally, Galileo Galilei, (1564-1642), took time off early in the 17 the Century to found telescopic astronomy, as something of a sideline to his monumental works on statics, mechanics, and hydrostatics. And as is well known, his advocacy of the heliocentric system landed him in house arrest outside Florence for the rest of his long life.

All these reflections are stimulated by the date assigned to the scientific Arabic manuscript, which straddles a period of intense astronomical development in the West, and stimulates a number of questions about the details of its contents. Is this a document describing a purely Arabic development? Is it based on observations, or does it merely recount material from some other source, e.g. a sort of textbook? If it is based on observations, are these naked eye or optical, and if optical what are the details? What picture of the planetary system does it espouse? If it is not of purely Arabic origin, what are its other sources, e.g. did the author know anything of the historical summary outlined above?

We have been unable to make progress with these documents, of which a mere translation may be inadequate to reveal all valuable insights. We have deposited a print of the De Witt Arabic manuscript photography at the Barker Historical Collection at the University of Texas at Austin.

## Mauritania Revisited

In the official report of the expedition published in the Astronomical Journal in 1976, Burton Jones records the follow-up expedition as having been on site at Chinguetti from November 6 to November 16. In fact the operations extended somewhat beyond this period of actual observations.

The personnel comprised Burton Jones himself, and Alassane Sy, the only ones to take part in the actual eclipse observations, Al and his wife, Marjorie Mikesell and the senior author of this book. The last named was honored with the title of leader of the team, though his status resembled that of the fabled leader reputed to have said, "I must hurry to catch them up, for I am their leader". This has suggested to the senior author that he should eschew the first person in this part of the work, and adopt a neutral designation, such as SA. After all, both Burton Jones and Mikesell had far more astrometric experience than he, Mikesell had had some considerable engineering, optical and photographic experience, Alassane Sy had a better command of French, more desert experience and far closer contacts with important people including members of his family in Mauritania, and Marjorie Mikesell was no supercargo, but a qualified scientist in her own right. SA seems to have been selected as the only permanent employee of the Texas Astronomy Department, with some knowledge of French, some African experience, and, because of his position in Austin, the only one qualified to act as paymaster in the field who could be held responsible for any errors of judgement that might be made.

It did not take long to show that taking in a small body of people to carry out a scientific task was an entirely different proposition from the eclipse itself, when there had always been the backing, if required, of the National Science Foundation and NCAR, with the availability of some items of equipment and even some strong arms of local laborers. The team did have the goodwill of Dr. Ba Bocar Alfa in Nouakchott, but certainly SA might have realised that the DeWitts had built up a fund of local amity which might have been called upon for help. As a result the feeling was that this team would have to do almost everything for itself, including one or two pieces of rather heavier engineering that one would not normally want to undertake without special tackle. For example, the eclipse team had had an overhead hoist at its disposal, and the follow-up team had not.

The involvement of SA began on October 29, when he took flight for Rochester, New York, home of much of the technical expertise of the Kodak Corporation, in close collaboration with the local University. He was carrying, not only his personal baggage, but items needed for the expedition, the whole totaling 103 lbs, which cost an extra $ 35 as far as Rochester. At Rochester he was met by the astronomer, Stuart Sharpless, and two representatives from Kodak. The object of this deviation was to pick up a new consignment of plates, meeting the same specifications as those previously supplied, namely coated on super flat glass, twelve inches square, and a quarter of an inch thick. There were 32 plates packed in boxes of eight. The total consignment weighed 62 pounds and was sent as unaccompanied baggage, addressed to SA at the offices of their friendly agent SOCOPAO in Dakar, originally chosen as the rendezvous for the team, except for Alassane Sy, who was to join in Nouakchott. SA was royally received, with a tour of the plate store, wine and dinner at the Faculty Club and a room at the Town House Motel. All this luxury was a trifle too much for the academic innards, which are recorded as somewhat disorganized.

Next day Dr. Berg came and the consignment box was opened and repacked, The day was overcast and fairly brisk, so SA took Sharpless and the Kodak scientists, Berg and Duthie to lunch, before heading for the airport, where a baggage excess to Africa of $ 571 was paid. Because of the low ceiling to New York, SA's flight was cancelled and he retreated to the International Hotel at PanAm's expense.

On October 31, after killing time, he left at 6.30 p.m. and then on to Dakar. The flight was smooth except for a certain amount of roughness off the African coast, which prompted the thought that this was the area where all the hurricanes which beset the Caribbean and the USA are born, originally from insignificant low pressure centers. Dakar was very misty, and after landing and a certain amount of argument at the Senegal customs desk, SA met the Mikesells and Burton Jones in the airport concourse. The delay experienced by SA had landed them on a public holiday, November 1, so they all repaired to their hotel, the Su-Nu-Gal, which was the scene of what might have been a seaside party at an expensive resort in metropolitan France. One formed the impression that, at least for the time being, nothing much had changed for the French colony from the days when they were the rulers of the country.

The hotel, as we know, had an inviting swimming pool into which his companions dived, but SA, not having thought that a visit to the Sahara might afford such an opportunity, had no swimming pants, so borrowed a a pair from a stock kept by the hotel. Unfortunately, when he dived gracefully into the pool these descended to his ankles, from which situation he beat a hasty retreat for cover. After lunch, the group took a taxi ride round town to the sea coast, the crafts market, the university and some other official buildings and the great mosque. All businesses were shut for the holidays, so the rest of the day was spent in relaxation, supper and sleep.

Dakar seemed to SA to resemble closely some other great ports, such as Durban or Rio de Janiero, but was probably more picturesque. There were very impressive large black ladies, decked out in brilliant robes, possibly for the national holiday, who wore on their heads brilliantly colored scarves in a sort of fly-away mode. They were often accompanied by menfolk, who were distinctly less conspicuous even though they might be crowned with a red tarbush. (Figure 83). On the beaches such ladies were to be seen cleaning enormous fishes evidently caught, (Figure 84), from the sea-going canoes pulled up on shore. These looked distinctly flimsy, with a prow structure seemingly hardly more robust that the fore canvas of a rowing eight, the main body of the vessel being a long boxy structure raised the best part of a foot, and painted with brilliant designs all over the outside. These were the apparently fragile vessels used by the local fishermen, but SA happened to know that they were excellent sea-going craft, having observed them from the deck of the South Africa-bound mail vessel a good many miles out to sea on several occasions.

The generality of the population apart from an occasional nomad glimpsed in the crowd in his desert dress, was rigged out in the now international norm of singlet and denims. The craft market was much the same as everywhere, with genuine indigenous crafts such as basket-making, utilitarian woodwork, and so forth, supplemented by toy-like carvings, bangles and other jewelry designed for the tourist trade. The Great Mosque was an awe-inspiring giant of ecclesiastical architecture, to be compared in size, perhaps wrongly, with St Paul's in London. The interior was a maze of arches, which, on inspection proved to be so arranged that every worshipper had a clear line of vision to the central area presumably occupied by the Imam at a prayer session.

On November 2, the team visited the offices of SOCOPAO where arrangements for travel to Nouakchott were made and SA was relieved to hear that his precious consignment of plates was already there. Much of the rest of the day was taken up with a boat trip to the island of Goree just a short ride from down-town. This was the notorious slave departure point where the wretches were loaded on to the ships of the Europeans for transit to Brazil, the Caribbean, or North America. It had by now acquired a deceptive atmosphere of floral calm,(Figure 85), with the exception of a number of large guns, evidently installed for protection of the port, but clearly severely damaged by shell-fire, possibly, one may speculate, as the result of the abortive attempt in 1940 of the Free French to take the city. Whatever may be the truth of this, it was also stated that the armament had had a kind of resurrection, serving as props in the making of the adventure classic movie, 'The Guns of Navarone'.(Figure 86). The guest list at the hotel was increased by the presence of a Russian volleyball team, who looked to have had a rough time against their local opponents. It was at this time that one first heard news of Eastern European factory ships at the fisheries, in this case Russian.

Next day, November 3, SA was in less than perfect form with a sore throat and suspicious bowels. However, the more serious medical news was that Marjorie Mikesell had tried a

Figure 83: On the beach at Dakar with deep-sea
fishing canoes.

Figure 84: The ladies gut a fish.

dip in the sea, and had encountered a sea urchin, which had presented her leg with a sample of its spines, which would have to be dug out. The party repaired to the airport and found that their craft, an elderly DC-4 was in need of attention, so that take-off was an hour late. The designation DC-4 has caused some discussion among the survivors, Bryce, an airman himself, insists that the planes used on the short runways in the various villages and towns, must have been DC-3 's, and this is undoubtedly the case, but here was a foreign flight from a capital to a capital, and it has been found true that craft designated DC-4 did exist. Indeed they apparently gave particularly good service during the Berlin air lift. The point seems to have been settled by Gerteiny (1967), who says that the fledgling air service had a fleet of DC-3's which could get into the rural airstrips and one DC-4 on charter from Spain. The flight occupied 1 1/2 hours over the sea with the plane exhibiting large flames from the exhausts. At Nouakchott there were many tiresome formalities. The party was met by Gorel and Dr. Ba, who helped with some kind of a contretemps at the hotel Mahraba, an incident mentioned, but in no detail, in SA's diary. SA had to borrow 2000 ougiya, designated as UM 2000, presumably as the result of a first contact with the new currency rules, which embodied a series of Catch 22's—transactions had to be in cash, currency could only be exchanged at a bank during its opening hours, it was forbidden to import or export any ougiyas. After a drink, some dinner, and a stroll in the dark along the main street, mostly strewn with sand, one was glad to get to bed.

Next day, November 4, SA went to the Ba's house with Gorel to await the man who was going to rent Land Rovers. SA had his usual trouble with identifying people and remembering names, but things gradually improved as the party sat on couches in the lounge, with shoes off, conversing with a large number of eminent local people, all of a very high level of intelligence and culture. Lunch was served upstairs

Figure 85: The peaceful charm of byways on
Goree belie the island's slavery past.

Figure 86: The coastal guns of World War II may
have been stand-ins in a movie epic.

with everybody sitting on a huge magnificent carpet.(Figure 87).Couscous came from a washbowl, followed by the usual ceremonial mint tea, and then a fish stew from another washbowl. After the meal, a richly carved silver ewer was brought, together with towels, for the guests to wash their hands, used bare during the consumption of the food. On the carpet, Marjorie's problem was expertly dealt with by Dr. Ba, (Figure 88). The party was then driven to the beach, which looked cold, bleak and dangerous, with cloud and mist. There was a jetty provided by the Chinese Government, with crabs and birds in attendance for what pickings were to be had. The party returned to Dr Ba's for dinner, and a meeting with the Land Rover man. By this time SA's interior was in a 'fair chaos'.

On November 5 SA's interior was no better, a situation which has prompted the thought in later life, that though the water-supply at Nouakchott was produced by the de-salinization of sea-water, perhaps other components were left in which ought to have been removed. The affliction, as recorded in his diary over several days does not seem to have affected his companions. Ironically, SA was the guardian of a box of medicaments, thought from experience to be most needed.

SA booked the homeward return flights of the party, hoping that he had correctly estimated the time requirements. A Land Rover was produced and after yet more time spent at the customs office, the air freight was actually released just before lunch. It was probably at this time that SA declared the value of the new plates at $ 11 per plate, assuming that they would be treated as consumable stores. With so little freight to be dealt with, the customs service could spend immense effort on minor matters, and this decision was to have awkward consequences later. To avoid having to carry large wads of local currency, since there was no means of encashment at Chinguetti, SA struck a deal at the government tourist office in Nouakchott, whereby he paid a large deposit on the use of the gîte to be opened there for the party's accommodation, with the understanding that any remainder would be paid after the return to the capital. The party then called on the US Ambassador, who received them warmly and gave them lunch, even including their Limey leader, with his diminished appetite. More time was spent on further organization of the Land Rovers, which seemed to be in very poor condition, in organizing a truck to bring down the equipment from Chinguetti, and arranging for a supply of gasoline. Exhausted by dinner time, SA settled the hotel bill and fell into bed.

On November 6, SA was still in trouble, and departure from Nouakchott was greatly delayed due to the inadequacies of the drivers—bad tires, no fan belt and the like. The road to the fairly important town of Akjoujt was good moderate black-top, running through almost deserted country, but there were camels, birds and occasional nomad tents to be seen in an otherwise forbidding landscape. Akjoujt was a town with a mine, operated by a French company, where there was a good French restaurant and a good French lunch for those with appetites.(Figure 89). Beyond Akjoujt the road degenerated into a track in quite reasonable condition, except that the transition from road to track lay over a very bumpy area with what seemed like odd piles of excavated dirt. As the vehicle banged and crashed its way over this intermediate section, SA, terror struck that his precious plates might be reduced to shards, insisted that the driver slow down even though this created the opposite risk that the vehicle might get stuck in the sand. On the track, the driver of the Land Rover in which SA was riding drove on the wrong side of the road, close up behind the lead vehicle. This prompted SA to the protest "Vous roulez trop proche. Vous risquez de recevoir une pierre dans la gueule." (You are following too close. You are liable to get a stone in your face). Hardly had this elegant piece of French prose left SA's lips, than a nasty sharp-edged chunk of very hard rock came sailing through the air from the

Figure 87: At Dr Ba's house,
lunch on a magnificent carpet.

Figure 88: Marjorie's sea-urchin spines are
expertly removed.

rear wheel of the lead vehicle, and reduced the windshield to smithereens. SA had to convince them that the best temporary solution was to knock out all the fragments and to close up all the rear windows, thus producing a pressure block of retained air which would help to exclude dust. However inadequate he was on a stone track, the driver proved to be extremely good on sand and at one stage when the road disappeared altogether, he was able to pick up tracks in the sand. At one point Al Mikesell insisted on a tire change,(Figure 90), but the dud tire reappeared, the good one not being on a wheel. Later the road entered mountainous country with immense fields of slaty rock. After a stop at a small village, the way led 40 km up a pass to Atar, where the Land Rovers were gassed up and a large barrel filled for use at Chinguetti. If recollection serves, it was here that the road passed through an immense field of broken bottles, and that the police stopped the convoy with some criticism, in vigorous Arabic, concerning the insurance of rented Land Rovers. All the way along, there had been stops for the drivers to say their prayers at appropriate times. After Atar the track became even poorer, and there was a flat tire. In the dark of a moonlit night, not much could be seen of the impressive Amogjar pass climbing up to the higher level of Chinguetti. SA had been very cross with the drivers when he discovered that the jerry cans used for water had not been filled, and at some time during the night they were stopped by nomads who were given a water-filled goatskin. In the end Chinguetti was reached and beds in the gîte, which had been specially opened for the party were more than welcome.

The morning of November 7 revealed the Land Rovers parked in the courtyard of the gîte, with the drivers hard at work trying to render their much abused tires fit for the return to Nouakchott. In due course they left with orders to pick up the team when they hoped their work would be done. (Figure 91). In the distance could be seen the lonely hut of their destination (Figure 92).There, everything seemed to be in order

Figure 89: The mosque at Akjoujt.

Figure 90: Desert debate on the state of a tire.

under the care of the guard, Bah Ould Soueidi, who had received the not-very-princely sum of the equivalent of $ 625 for his vigilance. The hut was opened with remarkable enthusiasm by Al Mikesell (Figure 93). For his part SA was feeling low, entertaining a thought that his well-known capacity for enterprise had got him into real trouble at last, and next day, November 8, he was no less glum because the weather was cloudy, and he thought that he might be setting a new record of a clear sky for an eclipse, followed by a failure of the follow-up on account of the weather. In the evening they were entertained to dinner by the Préfet, though it was the cooks from the gîte who cooked the whole sheep, and the cost still came on the residence bill. No matter, it was an elegant gesture of international amity, with the guests reclining on mattresses, drinking soft drinks, and chewing bits of sheep. SA, in recognition of his exalted status as leader, got served first (ahead of Marjorie), and so received a particularly intractable bit of sheep leg. Stories which had been recollected, of ceremonial offers of sheep's eyes turned out not to be true, or at least not applicable on this occasion, if indeed there was any substance to them. Later, at work, a focus plate was attempted, and found badly fogged, but the matter could not be pursued as it was cloudy all night.

November 9 was a day which reduced the team to almost total inactivity. The sky was overcast, thick all night and spotting rain. Getting the telescope temperatures right was no great task. There was a walk round the village and a meeting with two German geophysicists who were there to consider the policy for the use of the local water supply. They were apparently reluctant to install any more pumps, more powerful than the local shadoufs or the solar-powered pump, (which looked with its tower for all the world like an English parish church), for fear that they might overpump the supply dry. SA restarted reading the book he had just finished, bought an almost complete set of current postage stamps, and began to consider a plan of action if the weather stayed poor.

Figure 91: At the gîte the drivers prepare their
vehicles for return to Nouakchott.

Figure 92: Our modest destination, the distant hut.
The old walls show how the sand encroaches.

Figure 93: Mikesell could hardly wait to open the hut.

Figure 94: A main street in the oasis.

The next day was much the same only enlivened by a slide show at the school. Finally on November 11 the sky cleared, and some plates were obtained which looked acceptable, but the next day brought trouble with the processing, even though more plates could be secured. These were following a rather complicated and detailed protocol to provide exposures of the eclipse and comparison regions under the same conditions as had obtained on eclipse day. On days like this the group could take walks round the oasis, visit palm groves, see what the geologists were up to, and lead large swarms of eager little boys, friendly but curious, who did not want to let the visitors out of their sight. The oasis comprised a number of streets of stone houses, (Figure 94), and, on the outskirts, many reed huts, occupied by black-clad women and their children, (Figure 95). The team went to the street of the mosque to present to the Imam some of the printed material developed from Cécile's manuscript photography, (Figure 96) and in due course met the Imam himself in the presence of Gorel and a member of the gîte staff, (Figure 97).

These strolls were also an opportunity to achieve some practical understanding of the formation of the desert dunes, beyond what was found in Bagnold's book on the subject.(Bagnold,1940). Close up to a medium-sized dune one could see that, even in the lightest airs, it was topped by an air layer, perhaps a few inches thick, in which the lightest sand grains were being gently carried forward and deposited at the leeward end of the dune. They seemed to have a life of their own, not actively hostile, but completely indifferent to human ambitions. Faced with an obstacle, a dune slowly gathers itself on the windward side until it is ready to start slipping over on the lee, where a down-draught is initiated until the obstacle disappears completely. One can understand Bryce's experience with ancient ruins now being uncovered, which suggested that a previous incarnation of the town would now rise again to replace the one now in process of obliteration.

November 10 was only memorable for another slide show at the school, but the next day was clear and in the night (Universal time date November 12), six plates were obtained on III-O emulsion, each of the eclipse and comparison fields, with the artificial star grid and step-wedges imprinted. The Moon was very bright and there was a good deal of dust and haze. The plates were almost perfectly positioned, but there was some trouble with the processing.

Six more plates of both regions were obtained on the next night, this time with the sector running on all of them, unlike those on the previous night. There was some cloud but not enough to interfere. The next night a plate of the Moon through cloud was tried as a possible comparison to the eclipsed Sun.

November 14, with rain and a solid overcast, was spent exploring the village, but the next day was an improvement. The local municipality produced some wonderfully accoutered camels, on which, by way of example the Mikesells gave a demonstration ride.(Figure 98). There was also a visit to the local silversmith, where, naturally accompanied by tea, examples of his skill were purchased, (Figure 99), with Gorel in western style sitting on a camel saddle cushion (Figure 100).

That night, a series of ordinary comparison plates, and one of the Moon, then at almost the same declination as the eclipsed Sun, was obtained. Some others demonstrating the effects of focus change were added. On this, and on the final night, November 16 local time, various plates including the fields of Delta Andromedae and Epsilon Piscium, and of the Pleiades and Hyades star clusters were secured. A final series of eclipse and comparison fields, brought the total number of plates exposed to 33, some on new stock, some on left-over stock from the eclipse expedition.

Figure 95: On the outskirts there were
grass huts in the sand.

Figure 96: Waiting near the mosque to meet the Imam.

Figure 97: The Imam with a member of the gîte
staff and Gorel.

Figure 98: Marjorie Mikesell rides a camel.

Figure 99: Tea at the house of the silversmith.

Figure 100: Gorel, in western dress,
sits on a camel saddle cushion.

On November 17 the instrument was dismounted, an enterprise which presented real problems in the absence of the facilities available to the team who had installed it. The problem was that there was no overhead lift. After the borrowed lens had been removed the instrument was turned so that the tube was horizontal close to the ground. This left the counterweights high and inaccessible and only really removable, one by one, by a strong individual firmly installed on an elevated secure position, which we did not have. The solution seemed to be to lash the counterweight in position with some rope that was available, to remove the tube and then to lower the counterweights round to a position where they could be taken off close to the ground. The tube was removed successfully and taken out of harm's way, but unfortunately the rope restraint snapped and the counterweight whirled round to the bottom position. SA still has nightmares at the thought of what might have happened if anybody had been in the path of this more-than-sledgehammer blow. Injuries inflicted would have been instantly fatally crushing, but it did not happen, the deity having looked after one first-class fool that day.

The next day we packed up, handed over the hut to the Préfet, and drank champagne with Mr. Anezin at the gîte. A Land Rover arrived, but no truck. On November 19, Gorel, Marjorie and SA rode in the Land Rover, leaving Mikesell and Burton to supervise the truck loading, after which they would ride down in the Peugeot. The Land Rover journey was uneventful, passing the hired truck on its way at the top of the Amogjar Pass. Finding the Hotel Mahraba fully booked, we put up in the Hotel du Parc.

Next day we moved to the Mahraba, and SA embarked on the large amount of left-over business that had to be done:—payments for the rest of the bill at the gîte, and for gas, some of which we may have had on credit. Then a courtesy visit to the US Ambassador and a visit to the market for SA to buy boubous for his teenage sons. There was still no sign of the truck or the remainder of the party. A problem arose over payment for the Land Rovers with the renters wanting extra money, especially for the windshield replacement. Since the original deal had been struck through the intervention of Gorel, Dr. Ba became involved, and was very helpful. SA was pretty firm for sticking to the original contract price, because he only had restricted funds at his disposal, and uncertain commitments yet to meet. Eventually after long and exhausting negotiations his view prevailed, largely because there was no alternative. After dinner, the missing members of the party turned up, having stopped overnight at Atar.

November 21 was also a business day, much occupied with the shipping agent, who was to send the bulk of the equipment back by ocean freight, while the lens and the exposed plates were to go as personal air baggage, expected to incur large excess charges. A problem arose about some of the plates, imported but not used, which had been left behind in Chinguetti, SA taking the view that unused and abandoned plates had no value, while the agent pointed out that an incoming customs declaration had stated a value. Had the surplus plates been broken in the presence of a customs official, (who would have had to come from Nouakchott)? Obviously not. SA recalls the agent, (a Frenchman), giving way to a little outburst of tears before finding a way out of this impasse. Meanwhile, Al Mikesell took charge of the lens, in a wooden box wrapped up in a rug that they had purchased in Nouakchott. For some reason they had lost track of the official documentation proving that it could be repatriated to the USA without paying customs dues. On arrival at Kennedy airport in New York, they did not declare it, and fortunately the official's attention became fixed on the boldly marked metal case provided for the expedition. It was demanded that this should be opened, and fortunately proved to be full of dirty laundry, which put an end to the matter. All this was in the future, and while still at Nouakchott, there were repayments of various sums to Al and Burton over the expenses of their journey down, hotel bills, UM 32000 for the truck, and finally, on November 22, excess baggage payments for everybody. There, SA had to resist the desires of the customs officials to examine his excess baggage, i.e., the exposed plates, and heaving a sigh of relief at what now looked like a clear journey, discovered that he now had 8300 ougiya, which he could neither export, nor change, since that was only possible in business hours at the bank back in the capital. He solved this when he encountered a business man with, whom they had had friendly relations, and simply gave him the money. It afterwards transpired that this had caused the gentleman some trouble with the exchange authorities, but he did cover it later by the production of an entirely plausible-looking bill for some service or other. This got lost in the noise of SA's final accounting of the follow-up expedition, (in six currencies) rendered to the University.

The aircraft to Dakar was a Caravelle of the Air Afrique line, which deposited SA, now on his own, at the Dakar airport to await the arrival of a PanAm transatlantic flight. He whiled away some of the several hours of waiting with the purchase of a red frock covered with golden machine embroidery, for his wife. It fitted, and indeed can still be worn. He also got into conversation of a sort with a Bulgarian, a crew member from one of the factory ships with their associated fleet of catchers, fishing off-shore. Finally, the PanAm plane arrived, and crossed to New York where SA found, it being the day after Thanksgiving, that there was no available economy seat on the Braniff plane to Austin. He up-graded his passage, a thing which could then be done without the prohibitive transfer charges of the present day, and duly arrived in Austin to be met by his family, who were glad to see him because none of the postal communications either way had been successful, and, at the Astronomy Department, deposited his precious cargo. He brought with him vivid memories of an area of Africa new to him, and souvenirs in the form of a Mauritanian tea-pot and a hat, conical, made of stiff straw and leather, and

Figure 101: A springy hat made of reeds served
as a crash helmet in the Land Rover.

Figure 102: The much treasured carpet.

looking more Asian than African, which had the cardinal property of providing a springy impact with the roof of a galloping Land Rover. (Figure 101). This was not quite his final sign-off from the project, since, he would also import a much-treasured Mauritanian carpet, (Figure 102), and, later on, would be required to carry a considerable batch of plates to Heathrow, arriving at 8.00 am. on December 22 1973 there to hand them over to Burton Jones, who would work on their reduction.

## The Needle in the Haystack

The objective of all this effort was the determination of a single number, lurking well-hidden in a whole haystack of other numbers and procedures, seemingly instituted just to make the whole thing more difficult. We can perhaps avoid serious confusion if we start at the beginning and look at each step in turn.

We begin with the Smithsonian Astrophysical Observatory Star Catalog, which lists a position, known as accurately as possible, determined by the combined efforts of generations of astronomers, for each of the stars imaged in the eclipse and comparison fields. To derive the actual positions at any one time observed from any location, requires a computation, in which the constants are accurately known, to produce a set of sky coordinates to be seen from Chinguetti at the times that the photographic plates were exposed. These positions would apply if there were no terrestrial atmosphere and they were not imaged by a telescope. As we have seen, the positions of the stars are changed by the passage of light through the Earth's atmosphere. Atmospheric refraction can be fairly closely modeled from a knowledge of the local meteorological parameters, but not perhaps with the highest precision, so recourse is had to observations of neighboring comparison fields, with presumably closely identical refraction effects, the altitude and the meteorological circumstances being the same. The star field is now imaged by the telescope onto a flat plate, the circumstances of the transformation being dominated by the temperature of the lens assembly, which directly affects the scale of the picture. This again is going to be checked by a study of the comparison field and by efforts to control the temperature of the lens. Otherwise, there is no particular difficulty in the well-known tedious task of passing back from the record on the plate to the situation on the sky, if refraction has been taken care of. In the present case refinements were added to the process. In the traditional way the plate measurer sees each star imaged as a tiny patch of reduced silver emulsion and, by eye, does his best to bisect it with the cross wires in his eye-piece, and then writes down a number derived from a linear scale and a vernier head which operates the screw positioning the plate. In the present case these plates were to be measured semi-automatically on special machines, the PDS at Austin, assigned to Fritz Benedict for another program, and kindly made available by him, and/or, the Galaxy at the Royal Greenwich Observatory in Britain. They located star images by computer indexing of the eye-piece position, and then examined each star image in detail, each a tiny patch of silver, spread by atmospheric scattering, to locate its true center of density, and then to print out the results and plate position of the relevant star. As a still further refinement, the program converted the observed density of the silver to an estimate of the relative intensity of light which had caused it, using the photometric calibration provided by the step-wedge and fly-spanker images at the edges of the plate. This could correct a small possible error in the measured position, because photographic emulsions do not record light linearly, being reluctant to begin recording very faint sources, and saturating by developing all available silver, on reaching a certain relatively high intensity. Finally, the grid of artificial stars had been imprinted. This was to check on any distortion of the emulsion, essentially a gelatine layer impregnated with chemicals, which, as all film manufacturers hope, will, after processing, dry to a relatively robust layer without differential shrinkage.

Subjecting the plates to these complex procedures produced, as was customary at the time, enormous long tapes of data record, which then had to be analyzed to find the elusive needle. Burton Jones spent a considerable time at Austin, but eventually favored the Galaxy machine, and only achieved a final reduction after he had returned to the Lick Observatory in California. To his dismay, he found an unexpected difficulty in reducing the plates. They had been, before exposure, stored in a container at the rear of the telescope, referred to as 'The Basket', and then put in position for exposure, the slide being drawn, and the plate forced by spring pressure on to studs which defined the focal plane. Unfortunately there were four of these, and the springs were not sufficiently strong to ensure good contact. In fact, proper kinematic engineering practice should have specified three points to define a plane, but in this case some plates were in contact with different members of the group of studs introducing, a very small but significant, focus change in different directions on different plates.

The final result for the gravitational shift inferred for a notional star passing close to the solar limb was 1.66 ± 0.18 arc seconds. This was based on data for 39 stars, some distant ones used for plate scale determination. This result was in some ways a disappointment, the relatively small number of stars imaged evidently being due to the less than perfect conditions at eclipse time. The size of the probable error was sufficient to exclude any possibility of checking the prediction of an alternative theory, (Brans & Dicke, 1967), which predicted a value some 8% less than Einstein's.

It is not difficult to see the major problem of determination of the Einstein shift in visible light. The eclipsed Sun is imaged on a star field which may be thought of as a series of concentric rings, the outermost of which, being of a relatively large area, is most likely to contain a goodly number of stars, all too far from the Sun to show any effect of gravitational bending of light. The most useful stars for this purpose lie in ever smaller circles, of rapidly diminishing area, each less and less likely to include stars. Worse still, nearer and nearer to the Sun and its increasingly bright corona, more and more stars are likely to be obliterated by the increasing background light. It is, in fact, a losing game. The more we need the data, the less of it there is. The Texas expeditions strained to achieve the utmost rigor and refinement, and shall we say, did well, but not quite as well as Eddington in 1919, who took far more risks, and was really a bit cavalier with his reductions and procedures. But he did have four bright stars very close to the Sun, such as nobody else has ever had, and that made all the difference. (Figures 103 and 104).

## An Historical Retrospect

The entire project seems to have had the paradoxical result of verifying the 1919 results rather than the 1973 ones, which at first sight might seem a disappointment for the late comers. But if the object of scientific investigation is to gain understanding, this has definitely been achieved, as we can easily see.

Eclipses of the Sun, of course, necessarily occur with the Sun in its accustomed path, the ecliptic, which encompasses a great circle on the sky.

We can think of this as occupying 360 degrees, with the Sun, impaled, occupying half a degree at any one time. In other words we can think that there are roughly 720 different slots in which an eclipse can take place.

Only one or two of these offer star fields with close-in relatively bright members, each one of which ought to count double, because it doesn't matter whether the Sun abuts the left or the right side of the little star cluster. So we can roughly

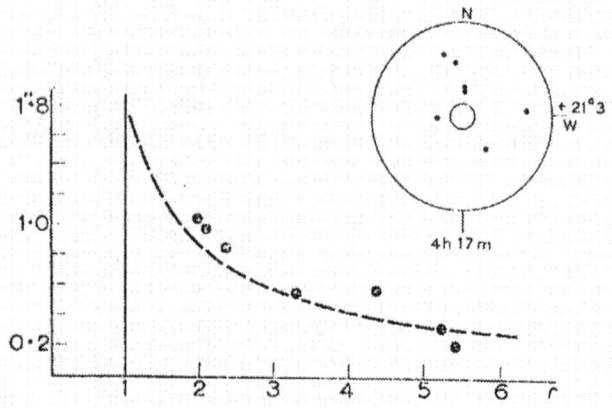

Figure 103: Von Klüber's plot of Eddington's 1919
observations. (From Von Klüber, 1960).

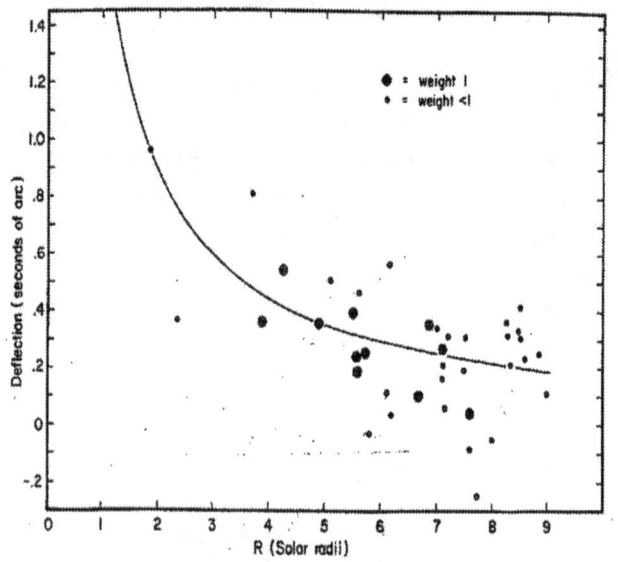

Figure 104: Burton Jones' plot of the 1973
observations. (From Jones, 1976).

say that, if, over the course of years, solar eclipses occur randomly round the ecliptic, there is a one in two hundred chance that an eclipse will occur in a favorable position for the gravitational displacement observations. There are a minimum of two solar eclipses in a year, and a maximum of five, but if the average is taken as three, this means that a favorable solar eclipse is only likely to occur every seventy or a hundred years, and if we add the additional conditions that the eclipse must be total and of long duration, this will make the chances of really favorable circumstances not merely centennial, but millennial.

Whether Eddington thought along these lines one cannot tell, but he was undoubtedly aware that the 1919 eclipse was the chance of a lifetime and he was eager to take advantage of it.

Nature provided a critical opportunity, but so did human history when one comes to look at it. Nobody could predict with confidence when the 1914-1918 Great War would end, with civil conditions restored enough to free personnel, resources, and travel facilities to allow a field expedition to be launched. That this happened was a very near-run thing. There was also the subject matter of the whole enterprise. At the beginning of the war various scientists had asserted their belief that scientific exchange was a subject superior to the deplorable enmities of the human race, but this stance became increasingly difficult to maintain as the horrors of the conflict burgeoned, year after year. Germany was especially execrated for the introduction of gas warfare, and by 1918 possibly the major proportion of the population of the victorious nations detested anything to a which a German label could be attached.

Einstein had been appointed Director of the Kaiser Wilhelm Institute of Physics in Berlin in 1914, and was thus a highly respected holder of public office, though one might argue, not as well-known internationally as in his home sphere.

Whatever his personal reservations may have been, he would certainly have been labeled both at home and abroad as a good patriotic German. He was thus fortunate that his scientific legacy was preserved for a wider public by the efforts of the neutral Willem de Sitter, and the pacifist Eddington, the latter in a position of unique power to secure English language publication of his research results. The fact that Eddington could be asked, possibly with some personal nudging, to produce the Physical Society's report on Relativity in 1918, is quite remarkable in view of the mental climate of the time.

When the results of the 1919 expedition were published they received a good deal of press publicity, though it must be realised that such mechanisms which did exist were almost trivial by comparison with those of today. But they did excite a certain better-educated stratum of the world's population, including, significantly, the young Chandrasekhar. The paradoxes of this esoteric subject titillated the public mind, and Einstein's name became widely known, and he could become a respected traveler to what had been hostile nations.

It was this feature which, only a few years later, when the German government classified him as Jew, that enabled him as a lionised guru to make his way to the United States, where he would have the opportunity to influence public affairs, most notably as the chosen intervenor to warn of the possible development of nuclear weaponry. The eclipse result also had the effect of launching Eddington into what must now be rated as a cosmological wild goose chase. (Evans,1998).

It is a strange tale. The story of an eclipse is apt to seem nothing but an anti-climax. Years are spent in preparation. The action is all over in a few minutes and the results may not be known until the final judicial authority supplied with a mountain of data on the microscopic topography of the plates, is ready to issue his verdict. We cannot expect the long term assessment of the 1973 results to match that of the first expedition, but it may be, if past experience is a guide, a little too soon to be sure.

## New Avenues of Research

The difficulties associated with making observations in visible light which, so far, have been restricted to times of solar eclipses, do not apply to observations made in, for example, radio wavelengths, where the Sun is not particularly bright, and strong radio sources can be observed close to the disk of the uneclipsed Sun. (Ohanian & Ruffini, 1994). The predictions can be pushed to a remarkable level of precision, because the parameters involved, the solar mass, the velocity of light and so on, are all known to a very great accuracy, and the observations bear this out.

It is assumed that visible light is no different from other wavelengths, and this is amply supported by modern observations of gravitational phenomena, such as, for example, lensing by a nearer massive galaxy creating multiple images of a more remote one. However, one does ask oneself whether there is any other situation in which gravitational deflection of visible light by the uneclipsed Sun might be detected. The one that occurs to the senior author, is based on his own observational experiences and a firm belief that, though special occasion observations are beguiling, with all the risks of accidents, there is a lot to be said for any humdrum procedure which might slowly produce the needed answer from a long run of observations, only trimmed by the usual exigencies of weather and other misfortunes. In this case it is the thought that occultations of stars by very young or very old phases of the Moon do so for starlight which has passed the Sun, it is true at a great distance, so that effects must be small, and might amount to some hundredths of a second of time. It would be hard to tease this systematic effect from the mountains of data available; but it should in fact be there. It is not a particularly inviting research project.

## Envoi

Will this experiment ever be done again? Almost certainly not. The days of adventurous expeditions to strange places to observe unique predictable astronomical events seem to have gone, to be replaced by the triumphs, and occasional unexpected failures, of space astronomy.

Was this expedition worth doing? Three decades ago quite a lot of highly qualified individuals thought it was. In the light of what we now know, it was not, but in the light of what we then knew, it was, and some of what we now know rests on the results of what we found out through this enterprise. Did we do it right? We tried for precise scientific discipline and rigorous conditions, perhaps failing to realize that some of the attempts at rigor got in the way of primary requirements. We would have been better off with a simpler lens, less liable to temperamental fits, but it is always a rule that in the field (or space?) simplest things are better. We had cooling equipment which depended on a supply of a rare commodity, namely water. We developed a considerable sympathy for the shortcomings of our predecessors after our own field experience. We would have been in far greater difficulties if we had not had the support of NCAR and their devoted staff to supply our material needs while the scientists got on with their business.

The enterprise, which called forth so much collaborative effort and good feeling in the Departments involved was a wonderful discipline and a great builder of morale. Does it also sound merely frivolous to say that it was all great fun and has left all the participants with happy memories?

It is time to say goodbye.

Many thanks to you all.

Austin, May 2002

# References

Bagnold, R.A., *The Physics of Blown Sand and Desert Dunes,* Methuen, London, 1941 & 1954

Baily, F., *On a Remarkable Phenomenon that occurs in Total and Annular Eclipses of the Sun,* Mem. R. Astron. Soc., **10**, 1, 1838

Beuchtel, E., *Niger,* Deutsche Afrika Gesellschaft, Die Länder Afrika, Band 38, Kwit Schroeder, Bonn, 1968.

Brans, C. & Dicke, R.H., *Mach's Principle and Relativistic Theory of Gravitation,* Phys. Rev., **124**, 925, 1961.

Buta, R., *Obituary of G.H. de Vaucouleurs,* Bull. Amer.Astron. Soc., **28**, 1449, 1996

De Vaucouleurs, G.H., *Obituary of Julien Péridier,* Q. J. R. Astron. Soc., **9**, 228, 1968

DeWitt, B., *Gravitational Deflection of Light: Solar Eclipse of June 30, 1973,* Nat. Geo. Soc. Research Reports, **14**, 149, 1973

DeWitt, B.S., Matzner, R.A., & Mikesell, A.H., *A Relativity Experiment Refurbished,* Sky & Tel.,**47**, 301, 1974

Douglas, A.V., *The Life of Arthur Stanley Eddington,* Nelson, London, 1956

Douglas, J.N., *Obituary of Harlan J. Smith,* Bull.Amer. Astron. Soc., **24**,1332, 1992

Dyson,F.W.,*Obituary of Andrew Ainslie Common,* Mon. Not. R.Astron.Soc.**64**,274,1904

Dyson, F.W., Eddington, A.S. and Davidson, C. *A Determination of the Deflection of Light by the Sun's Gravitational Field from Observations made at the Total Eclipse of May 29, 1919.* Proc.Roy.Soc. A, **220**, 579, 1920

Dyson, F.W. & Woolley, R.v.d. R., *Eclipses of the Sun and Moon*, Oxford University Press, 1937.

Eddington, A.S., *Space, Time and Gravitation*, Cambridge University Press, 1920.

Eddington, A.S.,*Obituary of E.T.Cottingham*, Mon.Not. Roy.Astron. Soc., **101**,131.1941

Evans, D. S., *Teach Yourself Astronomy*, English Universities Press, London, 1952—eds to 1970.

Evans, D. S. & Mulholland, D., *Big and Bright*, A History of McDonald Observatory, University of Texas Press, Austin, 1986

Evans, D.S., *The Eddington Enigma*, Xlibris, Princeton 1998

Fotheringham, J. K., *A Solution of Ancient Eclipses of the Sun,* Mon. Not. R. Astron. Soc., **81**, 104, 1920.

Fotheringham, J.K., *Historical Eclipses*, Halley Lecture, Oxford U. Press, 1921

Gerteiny, A.G., *Mauritania,* Frederick A. Praeger, New York, Washington, London, 1967

Gerteiny, A.G., *Historical Dictionary of Mauritania*, Scarecrow Press, Metuchen, N.J.& London, 1981

Gingerich, O., Oppolzer's *Kanon der Finsternisse*, translated from the German by, with a preface by Menzel, D.H. and Gingerich.O., Dover, New York, 1962

Jones,B.F., *Gravitational Deflection of Light* : Solar Eclipse of 30 June 1973 II. Plate Reductions. Astron. J., **81**, 455. 1976

Littman, M.,Willcox, K., & Espenak, F. *Totality* : Eclipses of the Sun, 2nd ed, Oxford University Press, New York & Oxford, 1999

Meadows, A. J., *Science and Controversy*, A Biography of Sir Norman Lockyer, The M.I.T. Press, Cambridge, Mass., 1972.

Mitchell, S.A., *Eclipses of the Sun*, Columbia University Press, New York,1923

Needham, J., Wang Ling, & de Solla Price, D.J., *Heavenly Clock: The Great Astronomical Clocks of Mediaeval China*, pub. in Assoc with Antiquarian Horological Society, Cambridge Univ. Press, 1960

Newton, R.R., *Secular Accelerations of the Earth and Moon*, Science, **166**, 825,1969.

Ohanian, H & Ruffini, R., *Gravitation and Spacetime*, W.W. Norton, New York and London, 1994, Ch 4, pp 194,195.

Texas Mauritanian Eclipse Team, *Gravitational Deflection of Light: Solar Eclipse of 30 June 1973, I : Description of Procedures and Results.* Astron.J.,**81**,452,1976

Unsigned, *Obituary of George Calver*, Mon.Not. R. Astron. Soc., **88**, 251, 1928

Unsigned, *Obituary of Albert Einstein*, Sky and Telescope, **14**,316, 1955

von Klüber, H., *The Determination of Einstein's Light Deflection in the Gravitational Field of the Sun*, Vistas in Astronomy, **3**,47, 1960. Pergamon Press, Oxford, Paris, New York, London.

Welton,K.C., Bland, P.A., Russell, S.S., Grady, M.M., Caffee, M.W., Masarik, J., Jull, A.J.T., Weber, H.W., Scultz, L., *Exposure age, terrestrial age, and pre-atmospheric radius of the Chinguetti mesosiderite. Not part of a much larger mass*, Meteoritics & Planetary Science, **36**, 939, 2001.